"The Object Lessons series achieves something very close to magic: the books take ordinary—even banal—objects and animate them with a rich history of invention, political struggle, science, and popular mythology. Filled with fascinating details and conveyed in sharp, accessible prose, the books make the everyday world come to life. Be warned: once you've read a few of these, you'll start walking around your house, picking up random objects, and musing aloud: 'I wonder what the story is behind this thing?'"

Steven Johnson, author of *Where Good Ideas Come From* and *How We Got to Now*

"Object Lessons describes themselves as 'short, beautiful books,' and to that, I'll say, amen. . . . If you read enough Object Lessons books, you'll fill your head with plenty of trivia to amaze and annoy your friends and loved ones—caution recommended on pontificating on the objects surrounding you. More importantly, though . . . they inspire us to take a second look at parts of the everyday that we've taken for granted. These are not so much lessons about the objects themselves, but opportunities for self-reflection and storytelling. They remind us that we are surrounded by a wondrous world, as long as we care to look."

John Warner, *The Chicago Tribune*

T0021675

Besides being beautiful little hand-sized objects themselves, showcasing exceptional writing, the wonder of these books is that they exist at all . . . Uniformly excellent, engaging, thought-provoking, and informative."

Jennifer Bort Yacovissi, *Washington Independent Review of Books*

. . . edifying and entertaining . . . perfect for slipping in a pocket and pulling out when life is on hold."

Sarah Murdoch, *Toronto Star*

For my money, Object Lessons is the most consistently interesting nonfiction book series in America."

Megan Volpert, *PopMatters*

Though short, at roughly 25,000 words apiece, these books are anything but slight."

Marina Benjamin, *New Statesman*

[W]itty, thought-provoking, and poetic . . . These little books are a page-flipper's dream."

John Timpane, *The Philadelphia Inquirer*

"The joy of the series, of reading *Remote Control*, *Golf Ball*, *Driver's License*, *Drone*, *Silence*, *Glass*, *Refrigerator*, *Hotel*, and *Waste* (more titles are listed as forthcoming) in quick succession, lies in encountering the various turns through which each of their authors has been put by his or her object. As for Benjamin, so for the authors of the series, the object predominates, sits squarely center stage, directs the action. The object decides the genre, the chronology, and the limits of the study. Accordingly, the author has to take her cue from the *thing* she chose or that chose her. The result is a wonderfully uneven series of books, each one a *thing* unto itself."

Julian Yates, *Los Angeles Review of Books*

"The Object Lessons series has a beautifully simple premise. Each book or essay centers on a specific object. This can be mundane or unexpected, humorous or politically timely. Whatever the subject, these descriptions reveal the rich worlds hidden under the surface of things."

Christine Ro, *Book Riot*

". . . a sensibility somewhere between Roland Barthes and Wes Anderson."

Simon Reynolds, author of *Retromania: Pop Culture's Addiction to Its Own Past*

"My favourite series of short pop culture books"

Zoomer magazine

# OBJECTLESSONS

A book series about the hidden lives of ordinary things.

*Series Editors:*

Ian Bogost and Christopher Schaberg

In association with

# BOOKS IN THE SERIES

# Air Conditioning

## HSUAN L. HSU

BLOOMSBURY ACADEMIC
NEW YORK · LONDON · OXFORD · NEW DELHI · SYDNEY

BLOOMSBURY ACADEMIC
Bloomsbury Publishing Inc
1385 Broadway, New York, NY 10018, USA
50 Bedford Square, London, WC1B 3DP, UK
29 Earlsfort Terrace, Dublin 2, Ireland

BLOOMSBURY, BLOOMSBURY ACADEMIC and the Diana logo are trademarks of
Bloomsbury Publishing Plc

First published in the United States of America 2024

Copyright © Hsuan L. Hsu, 2024

Cover design: Alice Marwick

For legal purposes the Acknowledgments on p. 149 constitute an extension of this copyright page.

All rights reserved. No part of this publication may be reproduced or transmitted in any form or
by any means, electronic or mechanical, including photocopying, recording, or any information
storage or retrieval system, without prior permission in writing from the publishers.

Bloomsbury Publishing Inc does not have any control over, or responsibility for, any third-party
websites referred to or in this book. All internet addresses given in this book were correct at the
time of going to press. The author and publisher regret any inconvenience caused if addresses
have changed or sites have ceased to exist, but can accept no responsibility for any such changes.

Whilst every effort has been made to locate copyright holders the publishers would be grateful to
hear from any person(s) not here acknowledged.

Library of Congress Cataloging-in-Publication Data

Names: Hsu, Hsuan L., 1976- author.
Title: Air conditioning / Hsuan L. Hsu.
Description: New York: Bloomsbury Academic, 2023. | Series: Object lessons | Includes
bibliographical references and index. | Summary: "Explores the surprising social, cultural, historical,
and environmental significance of air conditioning and of our efforts to control our climate"–
Provided by publisher.
Identifiers: LCCN 2023031662 (print) | LCCN 2023031663 (ebook) | ISBN 9781501377822
(paperback) | ISBN 9781501377839 (epub) | ISBN 9781501377846 (pdf) | ISBN 9781501377853
Subjects: LCSH: Air conditioning.
Classification: LCC TH7687 .H73 2023 (print) | LCC TH7687 (ebook) |
DDC 697.9/3–dc23/eng/20231003
LC record available at https://lccn.loc.gov/2023031662
LC ebook record available at https://lccn.loc.gov/2023031663

ISBN: PB: 978-1-5013-7782-2
ePDF: 978-1-5013-7784-6
eBook: 978-1-5013-7783-9

Series: Object Lessons

Typeset by Deanta Global Publishing Services, Chennai, India
Printed and bound in Great Britain.

To find out more about our authors and books visit www.bloomsbury.com and sign up
for our newsletters.

# CONTENTS

# INTRODUCTION

## WHY AIR CONDITIONING?

**FIG. 1** *The Hottest August* (dir. Brett Storey, 2019).

As I completed drafting this book in the unseasonably hot summer of 2022, air conditioning seemed to be everywhere I looked. Writing at a cafe patio in the sweltering heat (I was minimizing indoor time to reduce Covid-19 exposure), I was often engulfed by fumes from cars left running in the parking lot so the AC could continue uninterrupted while someone went inside to order drinks. In an Uber in Toronto, my driver pulled over and held his overheated phone up to the air conditioner until he could regain access to the itinerary. Amid record heat waves in Northern California, where I live, residents were asked to cut down on energy usage during peak hours so that countless AC units on full blast wouldn't overtax the grid. Along with the temperature, the cost of electricity soared, making AC increasingly inaccessible to lower-income households. In the news, climate activists glued themselves to paintings exhibited in air-conditioned museums. Teachers on strike in Columbus, Ohio included air-conditioned schools among their core demands. Fifty-three migrants from Mexico, Honduras, Guatemala, and El Salvador died of heat exposure and asphyxiation in a trailer truck abandoned in San Antonio, Texas, whose driver said he didn't notice the AC had stopped working.

Air conditioning is more widespread than ever in the 2020s, and we have never been more aware of its problems as a quiet, ubiquitous, yet unevenly distributed technology that requires massive amounts of fuel and creates greenhouse gas emissions. As climate change continues to devastate the most vulnerable communities worldwide, air conditioning has begun to lose its status as an iconic example of modernity's wondrous capacity to improve living conditions. It is now difficult to deny that air conditioning is an unsustainable and unevenly distributed technology that mitigates extreme heat in some places and in the present by emitting pollutants elsewhere, and far into the future. Air conditioning—a technology that middle-class Americans were long encouraged to take for granted—effectively transfers thermal comfort from the outdoors (including underground energy reserves and a stratosphere increasingly suffused with greenhouse gases) to the indoor spaces of those who can afford it. And as its emissions contribute to heating our planet, air conditioning makes itself increasingly indispensable: buildings, factories, digital devices, data servers, and life forms all depend on mechanical cooling to mitigate the effects of intensifying heat.

I was first moved to write this book after watching Brett Storey's haunting documentary, *The Hottest August* (2019). The film depicts a chilling disconnect between the omnipresent threat of climate change during one of New York's hottest summers on record and the everyday concerns of the diverse New Yorkers interviewed by Storey. For a film about climate change, there is surprisingly little mention of the topic—and yet its effects are everywhere. In the opening scene (Fig.1), for example, two union workers leaning out of an apartment window blame their workplace difficulties on immigrants willing to work for lower wages. Protruding from the window, their upper bodies mimic an AC condenser jutting out from an adjacent room. The air conditioner quietly entangles these workers with the patterns of immigration they see as an existential threat, since climate change (generated in part by AC emissions) has become a leading driver of migration worldwide. In scenes like this, the understated presence of air conditioning draws our attention to this nearly ubiquitous technology that allows many of the film's subjects (asked to talk about their anxieties concerning the future) to focus on economic concerns without directly addressing our immanent climate catastrophe. Wide exterior shots of walls interspersed with window AC units, along with the film's ambient soundtrack, create an uncanny sense of space and time: a present that can be comfortably inhabited only thanks to a technology whose emissions will linger long into the future. *The Hottest August* left me anxious about how air conditioning quietly desensitizes its users to the embodied effects of our hot summers, while its emissions make future summers even hotter.

What makes *The Hottest August* feel so unsettling—with both the climate and air conditioning units appearing only in the background—is the elemental intangibility of the changing atmosphere. Although our lives depend on it, air is notoriously difficult to represent. In both everyday speech and classical philosophical treatises, air serves as a metaphor for emptiness: we talk about things vanishing "into thin air," and Marx and Engels note that, under capitalism's waves of creative destruction, "all that is solid melts into air." As critic Tobias Menely explains, these metaphors imagine the air as an insubstantial and limitless repository: "To evaporate or to melt into air is not to disappear from view but to dissipate altogether. The atmosphere is a space so vast and empty that in it something, effectively, becomes nothing."[1] Menely argues that this idealization of air as that which is "outside" of history has made it difficult to understand how human history and activity have both transformed and been transformed by the materiality of air. Because it is designed to sustain ideal temperatures in a given space while emitting greenhouse gasses and waste heat outside, air conditioning perpetuates this problem of not being able to think about air in historical and material terms. In other words, air conditioning creates a separation between two kinds of space, and two kinds of air: interior air so comfortable that one doesn't have to give it a thought, and outdoor air that's supposedly so immaterial and plentiful that one also doesn't have to think much about it.

Air conditioning is an object lesson in both the power and the unacknowledged consequences of what we take for granted. It suffuses many of the spaces in which Americans—and, increasingly, middle-class people around the world—live, work, and play. It provides the luxury of not having to think about the weather, air quality, or how outdoor temperature is affecting one's body. As someone who spends a considerable amount of time reading, writing, and talking about books, I have often been a frequenter of climate-controlled spaces. During the years I lived in unairconditioned homes (and during my teenage years, when the thermostat was a constant bone of intergenerational

---

[1] Tobias Menely, "Anthropocene Air," *Minnesota Review* 83 (2014), 97.

contention at home—likely because my parents had grown up without AC in the subtropics), I have turned everything from Kmart patio displays, fast food restaurants, shopping malls, and the lobbies of hotels in which I couldn't afford to stay into makeshift workspaces. More recently, my work routines have been structured by regular stints in libraries, university offices, classrooms, and cafes. In such spaces, the writing and thinking body is sustained by the ever-present but hardly noticed comforts of air conditioning, heating, and ventilation. We might say we appreciate the atmosphere of a cafe, but we don't usually think about that atmosphere in material terms: the smell of brewing coffee, the ambient temperature, the filtered air. Air conditioning relieves us of having to think about the air, so that we can think about other things.

Writing in a cafe or library, my laptop is one of dozens scattered throughout the space, each thermally regulated by its own silent cooling system. Like our bodies, laptop computers function best within a certain temperature range. Everyone is working online, tapping into streams of data stored in distant servers that generate massive levels of heat. The internet data centers that house most of those servers consume vast amounts of energy, both to store data and to keep their storage units cool—usually around 65 to 80 degrees Fahrenheit. Here, too, air conditioning relieves the thinking body of its burdens: all the books and files we don't have to carry around because we can access them online, all the details we don't have to track down in a reference library or commit to memory thanks to Google. Perhaps, in the future, books like this one will be partially drafted—and "read"—by heat-intensive AI technologies navigating information drawn from continuously air-conditioned data centers.

## Descartes's Stove

Among the most eloquent examples of ambient temperature as a precondition for thought is the little-known story of Descartes's stove. (Because they were available long before artificial cooling, heating technologies are at the center of many early Western writings that reflect the cultural underpinnings of thermal comfort and air conditioning). In his *Discourse on Method*, published in 1637, the philosopher René Descartes recalls spending the night of November 10, 1619, alone in a quiet *poêle*—a room heated by a stove, to which he had retreated from the onset of winter. Here, he writes, "I was completely free to converse with myself about my thoughts." That night, Descartes had three dreams: in the first, a strong wind forced him against a church; in the second, he witnessed a light-filled vision in a room while a storm raged outside; and in the third, he sat at his desk and read in an encyclopedia *Quod vitae sectabor iter?* ["What road shall I follow in this life?"]. The young philosopher interpreted these dreams as a sign that he should devote his life to unifying the sciences, and he would subsequently attempt this by creating analytic geometry and by developing a philosophical method grounded in rationality—one that he believed would render us "masters and possessors of nature."[2]

Both of Descartes's innovations are steeped in dualism, or the belief that the mind is distinct from the body. Analytic geometry represents space as a vacuum oriented by abstract coordinates and occasionally inhabited by bodies, not as an atmosphere already filled with airborne materials of varying composition, density, temperature, and velocity. The stove-heated room offered Descartes a vision of thought isolated from climate: across his three dreams, he becomes increasingly sheltered from the storm outside as he comes closer to discovering his life's work. Only under conditions of climate control could Descartes begin creating a philosophy grounded in the fiction that the mind could be disentangled from the thinking, breathing body—a body sensitive to atmospheric factors like air quality and ambient temperature. This fiction is refuted by the etymology of "temperature": derived from the Latin for "mixture," *temperature* suggests a body that is blended with its surroundings. How would our understanding of Enlightenment philosophy and its legacies change if we revised Descartes's famous proposition, *I think, therefore I am*, to reflect the thermal conditions of a particular approach to thinking: *I think in a stove-heated*

---

[2]René Descartes, *Discourse on Method and Meditations on First Philosophy*, trans. Donald Cress, 4th ed. (Indianapolis: Hackett, 1998), 7, 35.

*room, therefore I am*? Or, perhaps, *I think my mind is distinct, therefore I am in a climate-controlled room*?

Why have philosophers had so little to say about Descartes's stove, and so much to say about his dreams, his resolve, and his conception of analytic geography on that winter's night? Suppressing the agency of the stove makes it easier to tell a simple story about the agency of the individual thinker. But it has made it that much harder to discern the subtle yet powerful ways in which modern air conditioning technologies condition thought, culture, and social experience. While researchers have devoted considerable attention to improving AC technology and studying optimal indoor temperatures, we know comparatively little about the implications of the variable availability of air conditioning across time and space for psychology, health, culture, and socioeconomic inequality. How do air quality and ambient temperature affect people's moods, relationships, and work experiences? Their life chances? What kinds of thoughts or feelings are enabled by comfortable temperatures, and what thoughts or feelings might be hindered by such comfort?

Descartes's suppressed dependence on the stove provokes this book's central question: how have social relations, culture, and everyday experiences been quietly shaped by air-conditioning technologies that modify indoor temperatures in some places while intensifying the effects of climate change elsewhere? This question approaches air conditioning not only as a technology for cooling the air, but more generally as a process that *conditions* human beings by conditioning the air we inhabit. While this book's reflections are oriented by the air conditioner as object, it also moves outward, following the consequences of air conditioners not only for the interiors they cool, but for what the novelist Mohsin Hamid calls "the great uncooled"—the vast populations worldwide with little or no access to cooled air, and all too much access to its environmental consequences.

## "The Ultimate Remoteness, Put-it-Somewhere-Else-Machine"

Climate control plays a very different role in the thinking of the environmental philosopher Val Plumwood than it did for Descartes. Best known for her visceral meditation on being attacked by an alligator, "Being Prey," Plumwood devoted much of her work to critiquing the host of ecological and social problems that stem from what she calls the "hyperseparation" of humanity from nature. Instead of taking for granted (as Descartes did) the conditions of comfort that support a specific mode of thinking, Plumwood takes the air conditioner's far-flung externalities as an occasion for ethical reflection. She coins the term "shadow places" to underscore how environmental externalities—or indirect side effects for which producers and consumers bear no legal or economic responsibility—are often concentrated in poor communities that have been made vulnerable by histories of racism and colonialism. Positioning Ogoni communities located at sites of oil extraction and pollution in the Niger Delta as a "shadow place" that enables climate-controlled rooms in the Global North, Plumwood writes, "We must smell a bit of wrecked Ogoniland in the exhaust fumes from the air-conditioner, the ultimate remoteness, put-it-somewhere-else-machine."[3] Instead of a technology that produces comfort, Plumwood reconceptualizes the air conditioner as a technology that draws comfort and well-being from one place and puts it in another. The air conditioner's capacity to generate—in one place—atmospheric conditions so comfortable as to go unnoticed comes at the cost of devastated air, water, and human and nonhuman lives in global sites of coal and oil extraction. The exhaust fumes created by air conditioning also displace thermal discomfort across space and time, contributing to greenhouse gases that are heating the planet in ways that disproportionately affect the most vulnerable places and populations. The consequences of generating comfortable air expand outward across a vast range of spaces that are especially susceptible to climatic instability.

One implication of this is that a book about air conditioning can't be restricted to the history of the technology, or to the perspectives of its consumers. As an object lesson, air conditioning stretches across toxic sites of fuel extraction, interior spaces suffused with machine-generated thermal

---

[3] Val Plumwood, "Shadow Places and the Politics of Dwelling," *Australian Humanities Review* 44 (2008), 139-50.

comfort, building exteriors warmed by "waste heat" from cooled interiors, and uneven climate crises across our warming planet. In sustaining comfortable spaces through ultimately unsustainable means, air conditioning produces a high-stakes differentiation of atmospheres—a society of sealed-off bubbles in which the quality of atmospheres, embodied experiences, and material objects come to depend on controlling the boundaries between internal and external spaces, as well as the boundaries between different kinds of interiors.

## Air Conditioning and Culture

We tend to think of the air conditioner as a machine for human comfort, but it has just as great an impact as an invisible infrastructure for industrial processes and cultural institutions. Aside from its effects on humans, air conditioning has an indispensable precondition for many nonhuman objects and materials, as well as ostensibly immaterial data: first invented in 1902 to control humidity that was disrupting the multicolor printing process at a Brooklyn printing company, air conditioning is now indispensable to everything from manufacturing, pharmaceuticals, and the food and beverage industry, to museum conservation and cloud data. "Process air conditioning"—or the modification of the atmosphere designed to optimize industrial processes—has made it possible to standardize and stabilize a vast range of materials, from baked goods, fermented foods, and chocolate, to fabrics, oil paintings, electronics, metals, and printed matter; whether you're reading this book electronically or in print, air conditioning helped make it possible. Laboratory science also relies on climate control to provide stable atmospheric conditions for controlled, repeatable experiments. If air conditioning makes it possible to imagine humans sealed off from their natural environment, it also sustains the fantasy that standardized and stable objects can be permanently sealed off from the effects of climate.

In both its effects on humans and nonhuman materials, air conditioning has become a powerful and insidious infrastructure of racial inequity—a tool for perpetuating racism on autopilot, by means of the everyday atmospheres we inhabit and transform. In addition to considering AC's ecological effects and its importance for nonhuman objects like museum displays and digitized archives, this book will dwell on the perspectives of people with limited access to air conditioning. By positioning climate control as a necessary condition for rational thought, Descartes marginalizes the thought and experience of people who—whether by choice or necessity—live amid uncomfortable or abnormal temperatures. In the ostensibly "public" climate-controlled spaces I discussed earlier—cafes, shopping malls, university buildings, libraries—people are differently vulnerable to profiling, harassment, criminalization, and forcible expulsion on the basis of race, class, gender, and/or disability. What kinds of embodied knowledge do we lose sight of when we assume that rationality requires thermal comfort? In addition to exposing the thermal inequities that sustain the modern world, we will see that dwelling outside of air-conditioned interiors can lead people to develop more sustainable and equitable techniques for inhabiting climates outside the temperate zone.

While this book will consider historical and scientific aspects of air conditioning—including its physical and psychological effects on humans—its main concern will be with the vast range of relationships between air conditioning and culture. While it will devote some attention to the growing demand and influence of AC in postcolonial nations such as Pakistan, Angola, and Samoa, this book's focus will be on the United States, which has the dubious distinction of being the leading nation in AC usage and energy consumption. American engineers, marketers, and architects have worked hard to create the culturally specific belief that a narrow, stable temperature range is a universal requirement of human comfort. At the same time, many cultural responses to air conditioning are unconscious or minimizing (like Descartes's passing mention of the stove), because air conditioning is most effective when it's least noticed. As the architect Juhani Pallasmaa remarks, we relate to atmospheric qualities like temperature and humidity through "unconscious and unfocused peripheral perception"—a mode of embodied sensation that precedes conscious thought.[4] Yet, despite this tendency to fade into

---

[4]Juhani Pallasmaa, "Space, Place, and Atmosphere: Emotion and Peripheral Perception in Architectural Experience," *Lebenswelt* 4:1 (2014), 243.

the background of our indoor experiences, air conditioning has become central to many of our assumptions about art, culture, thought, and the pursuit of knowledge—in part because air conditioning is a common feature of the laboratories, movie theaters, concert halls, libraries, bookstores, museums, and many (but not all) classrooms designed to house these pursuits.

In this book, I won't be telling you how to set your thermostat, or whether to stop frequenting air-conditioned spaces (for many of us, including myself, that would be highly impracticable). Nor will I make policy recommendations or dwell on the sustainable AC technologies of the future. Instead, I hope the histories, ideas, and cultural perspectives on AC explored in this book will shift how you think about temperature and its relationship to culture, the boundaries of the mind and body, and the supposedly non-negotiable nature of physical comfort. In addition to sharing information about the historical, ecological, and cultural implications of air conditioning, I hope to shift how you perceive this often-unnoticed infrastructure of (some people's) everyday life. Along the way, I'll also introduce projects—from conceptual artworks and public cooling centers to climatically appropriate architecture—that illustrate alternative ways of distributing of air conditioning and thermal comfort.

Atmosphere is an invisible backdrop of everyday life, and yet nothing is more consequential. The air conditioner teaches us not only that atmospheres are a product of culture, but also that culture is a product of its atmospheres. Because so many encounters with cultural objects that orient our senses of identity, beauty, and human meaning take place in climate-controlled conditions, air conditioning has come to permeate our sense of what a meaningful encounter or thought should feel like. While AC often takes the form of an invisible and anaesthetizing presence, we will take time to sit with works of art, literature, and film that endeavor to bring air conditioning into our awareness. My hope is that these cultural works can help loosen the grip air conditioning has on our sensory predispositions and our assumptions about "normal" conditions of life, and that they can push us to perceive and feel air conditioning differently. By considering both cultural representations of air conditioning and the AC technologies that condition many cultural experiences, I hope this book will help decenter the quietly privileged status of thermal comfort in contemporary culture.

# 1 THERMAL COMFORT

**FIG. 2** "This proves I should have married Jack Ditton!" Honeywell thermostat ad, *Life* 31:3 (Jul 16, 1951), 89.

Although air conditioning was first invented to condition industrial materials and printed matter, we have come to think of it primarily as a technology of thermal comfort. But "comfort" is a complex, culturally variable and contested term. What is more subjective than our sense of comfort? Whose comfort, and at whose expense? We tend to think of "comfort" in terms of ease and well-being, but it shares an etymological root—the Latin *fortis*, "strength"—with words like "fortitude" and "fortress." Comfort is not passive or innocent: it's a kind of power. It quietly calibrates people's capacities for feeling, thought, and action.

How did we come to think of temperature in terms of comfort, and to think of comfort in terms of a stable and unchanging temperature range? How did a relatively small temperature range become standardized as an ideal condition of "thermal comfort"? These transformations in our understanding and perception of temperature can be traced to efforts by air conditioning manufacturers to gain new markets for climate control technology over the course of the twentieth century. Even when it isn't sponsored by air conditioning engineers, research on thermal comfort often sets out to discover or test the idea that it increases productivity, while assuming that thermal discomfort is an incapacitating or unproductive state. What are the problems with comfort as a way of orienting our thermal experiences? And are we dismissing thermal discomfort too quickly, without pausing to consider what kinds of knowledge and activity it might enable?

# Comfort as Capacity

In the ensuing decades, air conditioning engineers quickly took up the idea of comfort as capacity, developing a marketing strategy focused on workplace efficiency. Even when they focused on marketing air conditioning customized to the needs of manufacturing processes, early engineers presented human comfort as a "happy by-product" for manufacturers. In many cases, air conditioning designed for manufacturing materials also improved environmental conditions for workers. Air conditioning marketers in the 1910s suggested that controlling factory humidity, temperature, and airflow would help attract workers, keep them satisfied, and enhance their productivity.[1] The idea that thermal comfort could influence human productivity and behavior emerged in connection with both empirical studies of correlations between temperature and work and racist fantasies about the supposed superiority of races originating in temperate climates (which I'll discuss in Chapter 5).

For the last century, labor capacity has been an important point of reference for research on thermal comfort. In the 1920s, scientists began conducting studies in sealed laboratories where test subjects could be observed

---

[1]Gail Cooper, *Air-Conditioning America: Engineers and the Controlled Environment, 1900-1960* (Baltimore: Johns Hopkins University Press, 2002), 56.

in different configurations of temperature, humidity, and air movement. Inside chambers such as the Harvard School of Public Health's air conditioning room, researchers instructed young white men to perform tasks such as "riding a stationary bicycle or lifting weights, to simulate labor."[2] Much more recently, the director of Cornell's Human Factors and Ergonomics Laboratory showed that employees made forty-four percent fewer typing errors and increased output by 150 percent when the temperature in an office was raised from sixty-eight degrees Fahrenheit to a more comfortable seventy-seven degrees.[3] Excess heat also affects people's capacity to work: drawing on data from over 58,000 factories in India, researchers found that output declines by two percent for every degree Celsius the annual temperature rises.[4]

But the effects of thermal comfort (and discomfort) extend far beyond labor productivity. Psychologists have introduced the "temperature-aggression hypothesis" to

[2]Michelle Murphy, *Sick Building Syndrome and the Problem of Uncertainty: Environmental Politics, Technoscience, and Women Workers* (Durham: Duke University Press, 2006), 21.

[3]Alan Hedge, "Linking Environmental Conditions to Productivity," slideshow based on presentation at the Eastern Ergonomics Conference and Exposition (New York, 2004), https://www.healthyheating.com/HH_Integrated_Design/Week%201/Linking%20Environmental%20Conditions%20to%20Productivity.pdf.

[4]E. Somanathan et al., "The Impact of Temperature on Productivity and Labor Supply," *Journal of Political Economy* 129:6 (June 2021), 1797-98.

account for apparent correlations between uncomfortable temperatures (especially heat) and hostile thoughts observed both in the laboratory and in aggregated media reports of violent incidents.[5] A recent study suggests that a significant portion of the "racial achievement gap" in schools can be attributed to obstructed learning on high-temperature school days, and to uneven access to air conditioning.[6] In addition to the well-documented health risks (such as heat exhaustion, cramps, heatstroke, respiratory and cardiovascular issues, kidney disorders, and pregnancy complications) associated with heat exposure, extreme heat generates ozone pollution associated with "asthma attacks, heart attacks, and other serious health impacts." Given the striking correlations between high temperatures, air pollution, and stillbirth, those who claim to value the lives of unborn babies should give some serious thought to climate justice.[7]

---

[5]Craig Anderson, "Temperature and Aggression: Ubiquitous Effects of Heat on Occurrence of Human Violence," *Psychological Bulletin* 106 (1989), 74-96.
[6]R. Jisung Park et al., "Heat and Learning," *American Economic Journal* 12:2 (May 2020), 306.
[7]Climate Central, "'Heat Islands' Cook U.S. Cities Faster Than Ever," *Scientific American* (Aug 2014), https://www.scientificamerican.com/article/heat -islands-cook-u-s-cities-faster-than-ever/. Christopher Flavelle, "Climate Change Tied to Pregnancy Risks, Affecting Black Mothers Most," *New York Times* (Jun 18, 2020), https://www.nytimes.com/2020/06/18/climate/ climate-change-pregnancy-study.html.

# Thermal Standards: Constructing a New Normal

Despite its powerful effects on conditioning people's health, mood, and actions, thermal comfort is rarely a topic of conversation. Because comfort is greatest when we notice it least, many people only become concerned about air conditioning when it's absent or malfunctioning. The unevenly distributed pleasures and privileges of air conditioning tend to go unnoticed for those who enjoy them the most. Air conditioning doesn't just make the air less noticeable—it also dampens our awareness of our bodies, and of the outdoors. We can become so acclimated to comfortable air that naturally occurring temperature variation comes to seem not only uncomfortable, but abnormal. Worse yet, because temperature preferences vary considerably according to people's "past thermal experiences and current thermal expectations," air-conditioned spaces *condition* people to associate comfort with increasingly cooler temperatures.[8] For those of us who are accustomed to being immersed in air-conditioned comfort, it can be difficult to recognize the problems and limitations of comfort.

---

[8]Richard de Dear and Gail Brager, "Developing an Adaptive Model of Thermal Comfort and Preference," *ASHRAE Transactions* 104:1 (1998), 145-67.

One problem with thermal comfort is that it is notoriously difficult to standardize. Early on, air conditioning (which works best when windows are kept closed) was frequently opposed by "open-air crusaders" who believed that comfort and health depended on access to fresh, outdoor air.[9] When the American Society of Heating and Ventilating Engineers (ASH&VE) published their "comfort zone chart" in 1922 to establish "fixed, rational standards" intended to demonstrate the superiority of artificial indoor air, the test subjects they relied on for thermal sensations were "predominantly white men employed (or seeking employment) in white-collar occupations," mostly dressed in suits and ties.[10] Despite this built-in experimental bias, the scientific authority of the chart (it was titled "Comfort Zone: Humans at Rest as Determined by Research Laboratory, American Society Heating and Ventilating Engineers") reduced thermal comfort to supposedly objective, "normal" standards of temperature and humidity. For example, the chart indicated that, at fifty percent relative humidity, comfortable ambient temperatures would range from sixty-seven to seventy-six degrees Fahrenheit. One year after their introduction, these experimentally established standards helped legitimate the decision to remove all references to "fresh air" from ASH&VE's ventilation code.

---

[9]Cooper, 64.
[10]Nicole Starosielski, *Media Hot and Cold* (Durham: Duke University Press, 2022), 54; see also Cooper 68-75.

While thermal comfort standards were subsequently revised and nuanced (for example, in 1966 the American Society of Heating, Refrigerating, and Air-Conditioning Engineers' Standard 55, *Thermal Environmental Conditions for Human Occupancy*, redefined thermal comfort in subjective terms as "the condition of mind that expresses satisfaction with the thermal environment"[11]), they continued to quietly center white middle-aged men's preferences as a universal norm well into the twenty-first century. Meanwhile, it is well documented that women tend to prefer higher indoor temperatures than men. Researchers have noted that standard thermostat settings in workplaces can create an uneven playing field that adversely affects women's capacity to focus on their work.[12] While gendered thermal disparities have received the most public attention—for example, in the viral College Humor sketch video, "Why Summer is Women's Winter" (2016)—the standardization of the thermal "comfort zone" also overlooks (or deliberately obscures) significant variances in people's thermal preferences resulting from cultural and individual metabolic differences.

---

[11]American Society of Heating, Refrigerating and Air-Conditioning Engineers, Standard 55-2017.
[12]Thomas Parkinson et al., "Overcooling of Offices Reveals Gender Inequity in Thermal Comfort," *Scientific Reports* 11 (2021), https://doi.org/10.1038/s41598-021-03121-1.

# Thermostat

Even if everyone could agree on a universally comfortable thermostat setting, the thermostat itself gives rise to further problems. As Nicole Starosielski writes in her groundbreaking study of thermal media, the thermostat's technical interface produced a "thermostatic subject, one that was supposed to desire stasis and continuity [and] avoid fluctuation." Drawing attention away from the diverse factors—such as physical activity, food, humidity, clothing, sun exposure, metabolism, and air circulation—that affect our sense of temperature, the thermostat encourages users to understand comfort solely in terms of ambient temperature. It detaches temperature from our interactive labor and energy (for example, adding fuel to the fire, or changing the AC settings manually) and transforms it into "a preference, a representation, and a consumable affect."[13]

As a normalized preference, thermal comfort (or, in some instances, a slightly uncomfortable coolness[14]) becomes associated with a range of ideological values, such as "civilization," freedom, hygiene, emotional moderation, self-discipline, health, domestic stability, and heteronormativity. Influential publications by climate determinists like Ellsworth Huntington, S.C. GilFillan, and Sydney Markham

---

[13]Starosielski, 33, 36-45.
[14]For more on white supremacist assumptions about coldness, see Chapter 5.

popularized the racist and Eurocentric belief that cooler, more stable climates (and especially the avoidance of excessive heat) provide a vital condition for "great" civilizations and nations. The neurologist George Beard, for example, devoted considerable attention to the enervating effects of America's intemperate climate in his influential 1881 book, *American Nervousness*, writing that "No other civilized country, ancient or modern, has ever had such a climate as this."[15] According to such theories (which were often a reference point for AC advertisers), mitigating the nation's weather would be vital to ensuring America's prosperity, because "in most of the continental United States, the climate is extreme in all seasons by western European standards, fiercely cold in winter and scorching or oppressively humid in summer."[16]

Early advertisements for temperature regulators targeted women in magazines like *Good Housekeeping* and *Ladies' Home Journal*, warning that variable indoor temperatures could lead to frustrated husbands and overheated children. In one ad from 1896, a husband tells his wife that their home's Johnson Draft Regulator ensures "an even temperature, and that means an even temper. An even temper keeps me at home, and completes our domestic happiness."[17] A 1951 Honeywell ad

---

[15]George Beard, *American Nervousness: Its Causes and Consequences* (New York: Putnam's, 1881), 170.
[16]Marsha Ackermann, *Cool Comfort: America's Romance with Air-Conditioning* (Washington, D.C.: Smithsonian, 2002), 11.
[17]Starosielski 42-44, 41.

(Fig. 2) features an overheated wife in the middle of the summer scolding her husband for not being "more thoughtful of my comfort," like her former suitor Jack Ditton would have been. As the key to stable and healthy bodies and families, thermal comfort was positioned as indispensable for the nation's future.

But normalizing thermal comfort also made everything else—that is, non-air-conditioned spaces and those who inhabit them—seem abnormal. The National Weather Bureau's introduction of a "Discomfort Index" in 1959, for example, presented a putatively scientific tool for measuring how far local, outdoor conditions deviate from standards for "comfortable" indoor temperature and humidity. Instead of considering whether local climates might affect people's thermal preferences, the Discomfort Index assumes that a supposedly objective standard of indoor air is "normal," and suggests that climates that people have inhabited for millennia are inherently, universally uncomfortable. As air conditioning enabled the deodorization of middle-class spaces and bodies, people whose bodies exhibited signs of abnormal heat also came to be stigmatized for their sweat and body odor. If thermostats were advertised as a means to more stable emotions and domestic relationships, people who could not (or didn't wish to) access "thermostatic subjectivity" were implicitly presumed to have "hot" tempers (a word that shares a root with "temperature") and unstable homes. If thermal comfort was the key to American prosperity and civilization, then people without air conditioning had no part in the nation's imagined future.

# Beyond Comfort

For those with the privilege to enjoy it, thermal comfort does enable certain ways of feeling and being, and it appears to enhance certain kinds of productivity. But in addition to its apparent benefits, it's important to consider the limitations of comfort. What do we lose when we inhabit buildings designed to minimize our thermal sensations? Although it has been excluded from the five classical senses, thermoception—or the sense of temperature—has distinct sensory receptors and occurs independently of tactile stimuli. Its evolutionary function is to cue animals to adapt to surrounding temperatures through thermoregulation: for example, by shivering, sweating, or modifying physical activity. Thermal sensations are most acute when temperatures are extreme, or when they change dramatically. When we inhabit a comfortable temperature range, thermoception becomes dampened and fades into the background.

The architect and daylighting consultant Lisa Heschong draws an instructive analogy between thermoception and the sensual pleasures of food. "While eating is a basic physiological necessity," she writes, "no one would overlook the fact that it also plays a profound role in the cultural life of a people. A few tubes of an astronaut's nutritious goop are no substitute for a gourmet meal." Thermostatic air conditioning can disconnect us from culturally meaningful experiences of "thermal delight" involving different, varying sensations of heat and cold: for example, a brisk morning walk, a bracing swim, or a visit to the sauna. Heschong argues that a "steady-state" approach to air conditioning goes to great lengths to save people from having to adapt to different thermal situations. And yet, "in spite of the extra physiological effort required to adjust to thermal stimuli, people definitely seem to enjoy a range of temperatures."[18]

Air conditioning has become a default, pre-emptive mode of thermal regulation: it protects us from extreme temperatures before we even have to feel them, and without our having to decide on an appropriate response. Our sensory capacities are not entirely inside us—they're how we interface with the world around us. By inhabiting temperature ranges that are designed to be hardly noticeable, we tone down our sense of temperature. Through habit, we become desensitized. How does this anaesthetic experience of thermal uniformity change people's connections with their bodies and environments? Here it is helpful to recall Starosielski's point that the thermostat disconnects the experience of temperature from the experience of energy expenditure: the thermostat dulls our attention not just to thermal sensations, but also to the constant energy consumption and emissions that keep air conditioning infrastructures running. We can wake up in an already air-conditioned room, unaware of the temperature and humidity outside. "One does not ask where

---

[18]Lisa Heschong, *Thermal Delight in Architecture* (Cambridge: MIT Press, 1979), 17, 21.

comfort has come from once it has become a habit."[19] The weather outside becomes a secondhand concern, at worst an inconvenience that can be avoided by moving from one air-conditioned bubble to the next.`

What about the positive affordances of thermal *discomfort*? What does discomfort make possible as a condition of thought, feeling, action, and relation? For one thing, thermal discomfort offers embodied knowledge of variable temperatures. Without experiencing it in everyday contexts, people become disconnected from unwanted sensations of cold and heat—as well as how those sensations can be entangled with feelings of helplessness, alienation, frustration, or anger. Especially for the disproportionately poor, racialized, Indigenous, and Global South populations that have limited access to AC, perhaps frustration or anger would be a more appropriate response to the climate (and to the human systems that shape it) than the "even-tempered" mood supposedly generated by a well-set thermostat. What kinds of thinking are possible *outside* of a perfectly comfortable, climate-controlled room? What kinds of knowledge are held in the body, opened up by our often-unacknowledged sense of thermoception?

Defined by an unchanging temperature range rather than something that varies widely with context, thermal

---

[19]Peter Sloterdijk, *Foams,* trans. Wieland Hoban (Cambridge: MIT Press, 2016), 376.

comfort erodes people's "climatic intelligence"—Eva Horn's term for the diverse techniques (siestas, spicy foods, later meals, clothing choices, the timing of outdoor activities) that people have relied on for centuries to mitigate extreme temperatures.[20] With the gradual erosion of these less technologically "advanced" yet often more sustainable and equitable methods of coping with the uncomfortable effects of ambient temperature, people's tolerance for temperature variation declines. This, in turn, increases people's reliance on air conditioning as the most immediately effective means of achieving thermal comfort—despite the fact that AC's waste heat and greenhouse emissions increase planetary warming in the long term.

---

[20]Eva Horn, "Air Conditioning: Taming the Climate as a Dream of Civilization," in *Climates: Architectures and the Planetary Imaginary*, eds. James Graham et al. (New York: Columbia University Press, 2016), 241.

# 2 BUBBLES

# A PARTIAL TYPOLOGY

**FIG. 3** "Vancouver." Unsigned promotional video, Dinner with a View, 2020. https://www.dinnerwithaview.ca.

Dinner with a View, a popular dining event staged in several North American cities during the Covid-19 pandemic, offers diners an opportunity to "dine in your own private 'bubble.'" These climate-controlled geodesic domes located in "iconic" sites in cities like Toronto, Montreal, and San Diego turned the social "bubble"—a widely practiced means of minimizing disease transmission—into an Instagrammable dining trend. Dinner with a View promised far more than comfort—it offered enclosed ecologies complete with plants and lighting: "Our domes get their inspiration from different regions of the earth and are transformed into terrariums with distinct terrains. A terrarium is an elegant encapsulation of an ecosystem; a living environment captured in time."[1] In the midst of a cold Canadian winter, patrons could dine together in a warm dome filled with desert plants amid a field of snow.

The airborne transmission of Covid-19 has focused public attention on both social and architectural bubbles. But we have been living with bubbles all along, as buildings, HVAC infrastructures, filter masks, and social boundaries have quietly organized modern life into discrete atmospheric containers. The philosopher Peter Sloterdijk theorizes this amassing of discrete yet adjoining air pockets through the concepts of "bubbles" and "foam." His book, *Foams*, theorizes twentieth-century society as "human foams" or "agglomerations of bubbles." Sloterdijk's concept of the bubble draws attention to human life as an aerial phenomenon—to the fact that humans (and many non-humans, too) can live, breathe, feel, think, relate, and act only with the support of atmospheric conditions

---

[1] "About," Dinner With a View (2021), https://www.dinnerwithaview.ca/en/about.

(for example oxygen, a livable temperature, and habitable radiation levels).

For Sloterdijk, who describes himself as a "student of the air," air conditioning is not just a technology, but a new way of understanding humanity in ecological, atmospheric terms. *Air conditioning* has a double meaning: conditioning the air that in turn conditions human life. The *Oxford English Dictionary* defines "condition" not only as "Something that must exist or be present if something else is to be or take place," but also as a "position with reference to the grades of society." It can refer to character or disposition, but also (as a verb) to the process of forming character of disposition. In Latin, *condīcĕre* means to talk together or agree upon something. What do people agree on without necessarily being conscious of it, simply by sharing the same atmospheric conditions?

Sloterdijk suggests that modernity can be dated to the use of mustard gas in the trenches of World War I—a weapon that targeted not human bodies, but the air that they depend on. Like these atmospheric weapons, twentieth-century engineers and designers reconceptualized humans as thermoceptive, olfactory, and breathing beings entangled with the air around them. Contrasting air-conditioned microclimates with Edmund Husserl's concept of a universal experiential "lifeworld," Sloterdijk writes: "Where there was 'lifeworld,' there must now be air conditioning technology."[2]

---

[2]Sloterdijk, *Foams*, 66.

In other words, philosophical meditations on human existence are no longer relevant unless they account for the qualities (including temperature, odor, and air pressure) of the different kinds of air that surround and sustain us, as well as the infrastructures that maintain those atmospheric conditions.

To illustrate how human bubbles coexist, Sloterdijk turns to the metaphor of foam. In foam, each bubble is surrounded by neighboring bubbles, separated by walls that both isolate and interconnect them. Twenty-first century buildings—cities, skyscrapers, shopping malls—are as foamy as Dinner with a View's assortment of air-conditioned domes. We live, work, and shop in different microclimates that are more or less sealed off from each other. If the distinctive atmospheric conditions of our bubbles influence us on multiple levels, then inhabitants of different bubbles are in some ways "inaccessible to the differently minded, differently enclosed, and differently-air conditioned."[3] Although communication between different bubbles may be difficult, it is also unavoidable, because they share walls and influence one another in unpredictable ways (for example, when air conditioning waste heat from one bubble affects the bubbles around it, or when electrical grids are strained by AC usage during heat waves).

---

[3]Sloterdijk, *Foams,* 173.

# Movie Theaters

The history of air conditioning's widespread adoption in America (and, increasingly, across the globe) can be traced across a range of bubbles, including factories, schools, laboratories, museums, movie theaters, shopping malls, casinos, office buildings, automobiles, and homes. Looking more closely at just a few of these structures will demonstrate how profoundly air conditioning has transformed the everyday spaces where we live, work, and play. From its early deployment in manufacturing processes, air conditioning soon spread to spaces of public leisure and consumption. In cities like Chicago, New York, and Los Angeles, movie palaces introduced air conditioning to lure audiences out of summer heat. When air conditioning was installed at Balaban and Katz's Central Park Theater in Chicago in 1917, it was a luxury that indicated the city's modernity and wealth. Widely adopted by movie theaters in the 1920s and 30s, AC established conditions for both the "summer blockbuster" and for a particular kind of viewing experience. Crowded theaters would no longer have to be suffused with stifling heat or the smell of sweaty bodies. As Marsha Ackermann writes, "The sensory environment that helped make movies wildly popular included not just moving images and, later, sound, but a cool darkness that allowed the rapt viewer to forget the time of day or the season of the year."[4]

---

[4] Ackermann, 47.

Along with the newsreel and the smell of popcorn, AC became essential to the otherworldly, escapist appeal of moviegoing—at least for audiences who could access downtown movie theaters under conditions of racial segregation. Many white, middle-class Americans first encountered AC at the movies, as they consumed classical Hollywood films that have been widely criticized for normalizing white supremacist, hetero-patriarchal, ableist, and nationalist beliefs. Given that thermal comfort has been shown to make people more receptive to ideas, AC may have been an important catalyst for the ideological conditioning conveyed by many classic Hollywood films. Billy Wilder's *The Seven Year Itch* (1955) exploits the equation of moviegoing with AC by dramatizing how a middle-aged "summer bachelor" (a man whose wife and child are staying at a mountain resort to avoid the heat) leverages his air-conditioned apartment to cozy up to a model desperate to get out of Manhattan's sweltering heat. The film's politically problematic content—for example, its objectification of native Lenape people and its reduction of lower-income women to potential playthings for well-heeled men—exemplifies the kinds of beliefs that went down easier when consumed in air-conditioned theaters.

But audience comfort was not the main objective of movie houses that installed expensive AC systems. Instead, their goal was to advertise cold as a consumer experience. With artificial displayed "icicles, icebergs, frost, and snow" prominently displayed alongside signs promising temperatures "20 degrees Cooler Inside," movie houses packaged cooled air as a technological novelty. What mattered was not the comfort of sitting audiences, but the curb appeal of cold breezes: "Many theaters were designed with a system that provided a positive pressure in the auditorium, forcing cold air out through the lobby and onto the street to tempt passers-by. Engineers called this 'advertising air.'"[5] Nicole Starosielski coined the term "coldsploitation" to describe the presentation of cooled air and Arctic-themed decor in tandem with a 1920s boom in "northern films" (such as Robert Flaherty's 1922 classic, *Nanook of the North*) set in Alaska and other cold places. Coldsploitation, she argues, "simultaneously exploited the sense of cold through colonial and racist representations of the Arctic and Antarctic, new forms of advertising and publicity, and emerging technologies of cooling."[6]

## Consumer Atmospheres

Across the twentieth century, air conditioning became a defining (if increasingly unnoticed) aspect of American consumer spaces. This began in the 1910s and 20s, when it was introduced selectively in department stores—especially "bargain basements" where sweaty crowds and stifling

---

[5]Cooper, 105.
[6]Starosielski, 76.

air could make for an especially unpleasant shopping experience. But the connections between air quality and shopping were revolutionized in 1956 with the opening of America's first fully enclosed mall, the Southdale Center in Edina, Minnesota. Designed by Victor Gruen, the building's interior temperature was set to 75° F year-round. Climate control required sealing the mall off from its suburban surroundings, despite its effort to simulate an accessible and vibrant public space.

As similar malls spread across the US landscape over the ensuing decades, they offered up carefully controlled, well-ventilated air intended to attract potential customers and encourage them to spend more time browsing. However, while cool interior temperatures became associated with elite shopping spaces (as the *New York Times* documented across a range of department stores, "the more ritzy the establishment is trying to be, the colder the air-conditioning is kept"), a recent study has shown that warmer temperatures (up to about thirty-one degrees Celsius) correlate with more positive product valuations on the part of consumers.[7] While opinions may differ about the optimal temperature for retail stores, filtered air also provides a blank canvas for scent marketing techniques ranging from the smell of freshly baked foods to the bespoke scents that are now diffused

---

[7]Yonat Zwebner, Leonard Lee, and Jacob Goldenberg, "The Temperature Premium: Warm Temperatures Increase Product Valuation," *Journal of Consumer Psychology* 24:2 (2013), 251-59.

in stores like Abercrombie, Victoria's Secret, and Vitamin Shoppe. Pleasant or unpleasant scents have been shown to affect consumer responses like shopping time, favorable product perception, and purchase intent: according to one study, "ambient scent yields a 3% increase in expenditures for an average setting and a 23% increase for a most favorable condition."[8]

The comforting, carefully orchestrated atmosphere of the shopping mall is an important and often overlooked theme in George A. Romero's classic zombie film, *Dawn of the Dead* (1978). Shot at the Monroeville Mall in Pennsylvania, which had its own indoor ice skating rink (featured in the film's closing credits, where it's overrun with zombies), *Dawn* focuses on a small group of survivors barricaded in a suburban shopping mall. The survivors take advantage of several architectural features connected with the mall's air conditioning system, like the building's walled-off exterior and its hidden air ducts. But air conditioning, and the way it conditions human instincts, is also part of what makes the mall so appealing to the invading zombies. "What are they doing? Why do they come here?" asks one character. The response: "Some kind of instinct. Memory of what they used to do. This was an important place in their lives." Like shoppers who spend more in warm temperatures, Romero's

---

[8]Holger Roschk and Masoumeh Hosseinpour, "Pleasant Ambient Scents: A Meta-Analysis of Customer Responses and Situational Contingencies," *Journal of Marketing* 84:1 (2020), 125-45.

zombies are oriented by an addiction to thermal sensation: "they ... feed only on warm human flesh." In one scene, a floor exhibit advertising tract houses ("Fully electric, central air") suggests that the behavioral programming of the shopping mall extends into suburban American homes. *Dawn*'s depiction of zombies swarming the mall evokes the insidious power of air conditioning as a background condition for thoughtless consumption and automated comfort.

## Domestic Bubbles

Americans first encountered air conditioning in movie theaters, department stores, restaurants, and other spaces of consumption; it wasn't until the decades following World War II that home air conditioning became widespread. In 1954-55, the National Association of Home Builders (NAHB) and the University of Texas at Austin developed the experimental Austin Air-Conditioned Village in Allandale, Texas. The project consisted of 22 unique homes each equipped with a different central air conditioning system. Inhabited by real families observed by researchers and weekly visitors, these homes tested the economic feasibility of central air conditioning (as opposed to window units) for single-family residences. After the first year, the NAHB reported that families in the air-conditioned homes slept ten percent longer, took more daytime naps, took up more hobbies, spent more time at home, encountered less indoor dirt and dust, enjoyed improved social skills, and generally experienced better "psychological well-being." [9]

On the heels of this widely publicized experiment, central AC became a standard feature of newly constructed homes in the United States. While only two percent of US residences had air conditioning of any kind (in most cases window units) in 1955, about twelve percent were air conditioned by 1962; by 1980, more than half of US residences were air conditioned.[10] New home construction surged in the post-WWII period, fueled by low mortgages for returning GIs and massive demographic shifts that included population surges (made possible by affordable AC) in the South and Southwest.

As the historian Gail Cooper has shown, architects and engineers collaborated to make air conditioning a defining feature of newly constructed tract homes. For example, in 1956 Levitt & Sons, Inc. worked with Carrier to install Weathermaker central AC systems in a suburban tract of 702 homes in Levittown, PA. In new construction, the cost of installing air conditioning was partially offset by eliminating traditional architectural features like porches, strategically positioned windows, and moving sashes that provided

[9]Srdjan Jovanovic Weiss, "Better than Weather: The Austin Air-Conditioned Village," *Cabinet* (2001), https://www.cabinetmagazine.org/issues/3/jovanov icweiss.php.
[10]Jeff Biddle, "Explaining the Spread of Residential Air Conditioning, 1955-1980," *Explorations in Economic History* 45:4 (Sep 2008), 402-23.

passive mechanisms of thermal comfort. Cooper explains that "The design elements that provided cooling and ventilation were no longer needed now that those functions were to be mechanized. This attempt to economize by redesigning the house around air conditioning depended on the substitution of a mechanical system for architectural features that provided passive cooling and ventilation. Such new houses required air conditioning to make them comfortable."[11] Constructed from cheaper, lightweight materials and designed with little regard for the local climate, tract houses left the issue of thermal regulation (traditionally an essential function of architecture) up to the air conditioner. In many cases, people's dependence on home air conditioning is an effect of architectural design, not a personal or biological characteristic. AC has now become a standard design feature, built into the vast majority of new homes (all new single-family homes in the South and ninety percent of new single-family homes in New England, for example, according to the National Association of Home Builders) and widely taken for granted as a normal feature of American life.

Passive cooling encompasses a broad range of architectural structures and practical techniques for regulating temperature. AC displaced the strategic use of building materials, paint colors, harder furniture surfaces, functional windows (as opposed to sealed "picture windows," which

---

[11]Cooper, 153.

often let in solar heat), house orientation, and cross-ventilation to mitigate heat. In many newly built homes, the front porch—a common element of Southern homes derived from West African and Caribbean architectural influences—lost its position as a distinctive site of everyday life connecting families with communal space.[12] While the NAHB boasted that children at the Austin Air-Conditioned Village spent more time indoors at home, it is important to recall that the porch, the stoop, and children playing outdoors were once pivotal to community life: as Jane Jacobs recalls in *The Death and Life of Great American Cities,* evening was "the time of roller skates and stilts and tricycles, and games in the lee of the stoop. . . . They slop in puddles, write with chalk, jump rope, roller skate, shoot marbles, trot out their possessions, converse, trade cards, play stoop ball, walk stilts, decorate soap-box scooters, dismember old baby carriages, climb on railings, run up and down."[13] Where Jacobs describes the sidewalk as a site of sociality, creativity, and cooperative play, central air kept children at home, often watching television. Along with architectural mechanisms of passive cooling, social rituals of thermal regulation—for example, sharing a cold drink, relaxing on the porch, or going to the movies—were also eroded by the thermostat. Air conditioning did not

---

[12]Jocelyn Hazelwood Donton, *Swinging in Place: Porch Life in Southern Culture* (University of North Carolina Press, 2001), 58.
[13]Jane Jacobs, *The Death and Life of American Cities* (New York: Vintage, 1992), 52, 86.

just separate American households from nature (as Henry Miller wrote in *The Air-Conditioned Nightmare* (1945), "nowhere else in the world is the divorce between man and nature so complete" as in the United States)—it also separated Americans from each other by turning home life inward.

Two high-profile examples illustrate some of the implications of inhabiting homes that have been designed as hermetically sealed bubbles. Alison and Peter Smithson's influential 1956 design for the House of the Future had no exterior windows, and interacted with the outdoors primarily through a speaker, microphone, and electronic entry door. The state-of-the-art central heating and AC system created a purified atmosphere noticeably different from the outdoor air. At the center of the design was the fully enclosed courtyard, which offered the amenities of "nature" without requiring any interaction with the outside world. The House of the Future envisioned future homes as self-enclosed machines for living, where nature itself could be recreated—with some filtering and modification—in the heart of private space. The Smithsons's project exaggerated design trends such as suburban isolation and picture windows, envisioning home as a place where you literally breathed different air than people outside.

Decades after the Carrier Engineering Company installed central air-conditioning in the White House in 1930, President Richard Nixon developed a habit of running the AC and fireplace at the same time. The president did much of his speech-writing in the Lincoln Sitting Room, "with the air conditioner up high and a fire in the fireplace."[14] Nixon reportedly indulged this habit even during the 1973 oil crisis, when many Americans were suffering from energy shortages (during the extended energy crisis, President Jimmy Carter would later ask Americans to turn down their thermostats to 65 degrees or less in winter, and no lower than 78 in summer). Like the House of the Future, Nixon's peculiar practice of thermal comfort exemplifies how air conditioning can cut people's home life off from social and environmental conditions that are inescapable for those outside the bubble.

Remade as domestic bubbles, homes offered new capacities for work, leisure, and display. As the architect and researcher Srdjan Jovanovic Weiss notes, "The reports from women at the Austin Air-Conditioned Village testifying to less dirt and dust in the house in turn translated into a language of visual purity. By virtue of being separated from outside dirt, white colors in bedrooms and living rooms became affordable. Previously seen as luxuries on the movie screen, white rugs, curtains, and upholstery became assets for a 'perfect' bedroom, only available with the purchase of the air-conditioned house."[15] If the suburbs were in many cases destination points for "white flight" from the race and class proximities of the city, AC enabled families

---

[14]Jay Mathews, "Nixon Revisited," *Washington Post* (June 14 1990), https://www.washingtonpost.com/archive/lifestyle/1990/06/14/nixon-revisited/73a34849-3764-4d89-96f2-37bd3ea93297/.
[15]Weiss, "Better than Weather."

literally to immerse themselves in whiteness. Cooler indoor temperatures also created conditions for more time spent at home and, thus, more time spent doing domestic labor: not only did these white rugs, curtains, and upholstery have to be carefully maintained, but "Women in the Austin Air Conditioned Village baked and cooked more than in previous summers, serving their families more hot food."[16]

## Skyscrapers

Office buildings and other high-rise structures have also been remade as bubbles. Air conditioning facilitated a transition from "sprawling H-, T-, and L-shaped buildings that ensured a high proportion of window exposure to interior space" to more space-efficient block-shaped office buildings.[17] By 1962, half of all US office buildings were air-conditioned.[18] As with family homes, passive cooling and ventilation features like ventilation shafts and outside exposures were displaced by more economical designs that presumed dependence on air conditioning. Office buildings like the United Nations Secretariat (completed in 1951) and New York's Lever House (completed 1954) could incorporate floor-to-ceiling glass as a building material, trusting that the AC would offset the

---

[16]Cooper, 174.
[17]Ibid., 160-63.
[18]Weiss, "Better than Weather."

resulting increase in heat. Glass structures may have looked clean and futuristic, but they were not always energy efficient: the iconic glass walls of the UN Secretariat "increased its air-conditioning requirements by 50 percent."[19] Along with air conditioning, glass-and-steel design enabled architects to bring air circulation and natural light to the stale "dead space" in the center of block-shaped buildings. Like the transparent surface of a bubble, the glass wall afforded visual transparency coupled with atmospheric separation. The contemporary high-rise—a prominent symbol of urban wealth, power, and modernity—would not be possible without the thermal comfort, ventilation, and sealed glass designs made habitable by air conditioning. Among other things, this means that—like modern homes designed with minimal attention to passive cooling techniques—sealed office buildings are especially vulnerable to power outages during heat waves.

## Individualized Bubbles

As we have seen, the twentieth century saw air conditioning spread to a range of large, institutional bubbles such as factories, schools, hospitals, government buildings, shopping malls, office buildings. The 2010s saw a shift in the opposite

---

[19]Cooper, 163.

direction, towards individually tailored air-conditioning. In an effort to address people's different thermal preferences while increasing energy efficiency, apps such as Comfy and UC Davis's TherMOOstat enabled building users to report their own thermal sensations to building managers. The Nest Learning Thermostat, first introduced in 2011, adapts to varying thermal preferences in different rooms, and at different times of day. RoCo (the name stands for "roving comforter") is a robot developed at the University of Maryland to follow its user around, sense their body temperature, and provide cooling calibrated to a user's particular preferences and immediate surroundings. These thermal feedback technologies shift from the idea of a standardized comfort zone to personalized processes of "thermal care."[20] By gathering and responding to individual temperature preferences, these technologies shift the scale of comfort from the building to the individual: a specific bedroom, cubicle, or building zone. A similar trend can be seen in recently introduced "wearable AC" products like Sony's Reon Pocket 2 and Blaux's Wearable AC Plus. Worn inside the clothing or around the back of a user's neck, these micro-ACs provide cooling (Blaux's version also provides filtering and air ionization) scaled to individual bodies and preferences. While individually tailored climate control technologies may be more sustainable and comfortable

---

[20]Starosielski, 61-69.

for individual users, they also deepen our conscious and unconscious attachments to air conditioning. The dream embodied by these products is to enclose individual users in personalized thermal bubbles, immunizing them against extreme temperatures and pollutants in the air around them.

## Dystopian Bubbles

When we recognize that middle-class Americans have already been living, working, and playing in air-conditioned buildings for decades, the more explicit bubble-spaces of science fiction no longer seem like the stuff of the future. While architectural bubbles have often been imagined (and sometimes realized) as structures that would enable humans to colonize and thrive in uninhabitable climates on Earth and in outer space, they are also a necessary and deeply ironic tool of survival in dystopian stories like Hugh Howey's "Wool" (2011), Bong Joon-ho's *Snowpiercer* (2013), Paolo Bacigalupi's *Water Knife* (2015), and Prayaag Akbar's *Leila* (2017). Bacigalupi's novel, for example, focuses on water conflicts in a near-future US Southwest where the wealthy have retreated to self-sufficient Arcologies sealed off from the extreme heat, ubiquitous dust, and brutal drought outside. Despite its focus on the theft and privatization of water, *The Water Knife* also dwells on air as a medium of life, comfort, and suffering. In one scene, a character feels "cocooned from the world outside" in a Tesla where "cool A/C [pumps] in a steady hiss through HEPA filters"; in another, Arcology residents shielded from the climate by insulated glass, AC, and top-notch air filters appear to be ignorant that a dust storm is wreaking havoc outdoors. In any case, they "might not even care that the world was falling apart outside their windows."[21] In Akbar's *Leila*, waste heat emitted by an elite domed community's AC system sets the homes of people living outside on fire. The irony in such dystopias is that air-conditioned bubbles allow some people to continue living comfortably on a planet that air-conditioning has helped make uninhabitable. As the Situationist International put it in a 1962 commentary on Cold War bunker mentality, "The world of shelters acknowledges itself as an *air-conditioned vale of tears*."[22] Extrapolating from the increasingly sealed-off spaces in which we're already living and working, these dystopian visions offer cautionary tales about the social problems—health disparities, entrenched inequalities, exploitation, and indifference to suffering—that are inherent in the bubble as an architectural ideal.

---

[21]Paolo Bacigalupi, *The Water Knife* (London: Orbit), 121.
[22]"The Geopolitics of Hibernation," trans. Ken Knabb, in ed. Ken Knabb, *The Situationist International Anthology,* revised and expanded edition (Berkeley: Bureau of Public Secrets, 2006), 103.

# 3 WEATHERMAKING

## VICIOUS CYCLES

**FIG. 4** The weather project, 2003. Monofrequency lights, projection foil, haze machines, mirror foil, aluminium, scaffolding. 26.7 x 22.3 x 155.44 m. Installation view: Tate Modern, London, 2003. Photo: Ari Magg. Courtesy of the artist; neugerriemschneider, Berlin; Tanya Bonakdar Gallery, New York / Los Angeles. © 2003 Olafur Eliasson.

The inescapable yet frequently forgotten connections between air-conditioned bubbles and the planetary climate were at the heart of the Icelandic-Danish artist Olafur Eliasson's monumental Tate Modern Museum installation, *The weather project* (2003; Fig. 4). The work deployed monofrequency lights, haze machines, a semi-circular screen, and reflective materials to create an artificial sun immersed in mist. The reflective ceiling—where visitors could see themselves illuminated in orange light—revealed that the top half of the artificial sun was a reflection. By presenting the sun—the main driver of naturally occurring weather—as a manufactured image alongside visitors' own reflections, Eliasson blurred distinctions between outdoors and indoors, nature and industry, the weather and us. As one critic explains, the work's "importance rested on its obvious constructedness— since the pipes of the mist machines and the wires and lamps of the sun were all openly exhibited—the opposite of the near invisibility of climate-control equipment in a typical building."[1] *The weather project* was a site-specific installation, designed to be exhibited in Tate's expansive Turbine Hall (once filled with turbines powered by coal boilers, before the Bankside Power Station was reincarnated as one of Britain's most prestigious galleries). At once awe-inspiring

---

[1]Jim Drobnick, "Airchitecture: Guarded Breaths and the [Cough] Art of Ventilation," in *Art, History and the Senses 1830 to the Present,* eds. Patrizia Di Bello and Gabriel Koureas (London: Ashgate, 2017), 163.

and candid about its artifice, Eliasson's sun provoked visitors to consider the immense effort and fuel required to sustain artificial bubbles. The continuous burning of fossil fuels to maintain indoor environments creates artificial atmospheres both indoors (where the weather is now carefully designed) and *outdoors* (where weathermaking emissions are released without much care, and easily forgotten as they slowly accumulate out of sight). What does it mean to encounter the weather as a "project"—not just conditions we adapt to and endure, but a climate that we help make? How can we recognize our many weathermaking activities and approach them consciously, deliberately, and equitably?

# Weathermaking, or Moving Heat Around

The modern air conditioner, invented by Willis Haviland Carrier on July 17, 1902, consisted of a fan circulating air over steam coils filled with cool water. As humidity condensed onto the cold coils, the air became cooler. This design was refined by incorporating compressed refrigerant fluids, which absorb considerably more heat and moisture from the air as they expand. Most of today's air conditioners run on a similar but more elaborate design: air is blown over evaporator coils filled with cold refrigerant fluid (or fluid that typically evaporates at a temperature considerably lower than room temperature); as the refrigerant fluid absorbs heat from the air blown over it, it changes into a vapor. Next, the vapor moves through a condenser, where it releases its excess heat to the outdoors while condensing into a high-pressure liquid. The high-pressure refrigerant then passes through a metering device, where both its pressure and temperature drop sharply. "Waste heat" and excess humidity generated during this cycle are continuously released outdoors from the condenser unit. Air conditioners frequently also remove particulates from the air in the process of cooling it. It is evident from this design that air conditioning is not just about regulating temperature: as the AC historian Gail Cooper writes, the "four essential functions of modern air conditioning" are "temperature, humidity, cleanliness, and distribution of air."[2]

More fundamentally, Carrier's design illustrates that air conditioning is not a "cooling" technology so much as a technology for moving heat around. It moves unwanted heat (and the humidity that goes along with it) from some indoor spaces to the outdoors. Similarly, filters may clean the air, but dirty filters accumulate in distant landfills. And, at least before the recent, slow, and spotty adoption of solar energy, all of this machinery for cleaning and cooling interior air has run on fossil fuels that pollute and heat the planet.

When the Carrier Engineering Corporation named its first residential unit the Weathermaker in 1928, the company

---

[2] Cooper, 12.

had indoor weather in mind. The idea of making one's own indoor weather had both futuristic and godlike overtones. It promised to separate the spaces of everyday life from the outdoors, freeing users from both daily and seasonal fluctuations in temperature. In guaranteeing a stable indoor environment, the air conditioner detached home life from seasonal cycles that have organized human life and customs for generations. Seasonal weather was now something that happened entirely outdoors. You could temporarily tolerate it, visit it, or avoid it altogether. Although this original Weathermaker machine was designed to provide heating for the winter (or "winter air conditioning"), "Weathermaker" soon became one of the most common terms in HVAC branding.

## Externalities

But weathermaking has a dark side. Air conditioning maintains indoor "weather" conditions at great cost to the planetary environment. Although many buildings are now designed as insulated bubbles sealed off from their surroundings, architecture cannot overcome the laws of thermodynamics. An air-conditioned building will always tend towards equilibrium with surrounding air temperatures. And as millions of air conditioners release waste heat outdoors, they raise surrounding temperatures. Air conditioning emissions and waste heat are significant contributors to the "urban heat island" effect (which I'll return to in Chapter 4), which is further intensified by nonporous and reflective building materials designed to keep air-conditioned interiors sealed off from their surroundings. On a planetary scale, rising demand for indoor comfort has contributed to increasingly erratic outdoor weather. The climate-controlled bubble enacts a widespread tragedy of the commons, privatizing comfort while making the planetary climate less habitable.

Extreme heat resulting from climate change and the heat island effect can cause increased air pollution and ozone concentrations at ground level. Heat accelerates the chemical reactions that produce ozone, while also enabling wildfires (and smoke) to spread rapidly. Widespread AC usage during hot days also considerably increases airborne pollutants emitted by power plants. A recent study focusing on the eastern US estimated that increased air conditioning in response to climate change could account for about 654 annual deaths linked to fine particulate matter and 315 annual deaths linked to ozone.[3] In all these externalities, we can see an insidious vicious cycle at work. AC emissions intensify outdoor heat and pollution, thus increasing people's

---

[3]David Abel et al., "Air-Quality-Related Health Impacts from Climate Change and from Adaptation of Cooling Demand . . ."*PLOS Medicine* 15:7 (2018).

reliance on air conditioners and mechanical ventilation for cool, breathable air.

This calls to mind one of the most common self-help clichés: the surprising power of what we choose to do every day. As it sustains the enclosed bubbles in which some people live, work, play, and gather every day, air conditioning also accumulates atmospheric toxins and greenhouse gases outside those bubbles. According to the Energy Information Administration, AC equipment is used in eighty-seven percent of US homes, and heating and cooling account for about half of the average US household's annual energy consumption; in commercial buildings, the US Small Business Administration estimates that HVAC accounts for about forty percent of energy consumed. The Department of Energy notes that air conditioning alone in the US releases about 117 metric tons of carbon dioxide into the atmosphere. Climate control is important to many Americans' values and sense of comfort, and, unsurprisingly, the US has the highest per capita energy consumption in the world (China overtook the US as nation that consumes the most energy overall in 2009). Worldwide, according to the World Bank, cooling technologies account for up to ten percent of all greenhouse emissions, and this figure is expected to double by 2030.

Air conditioning equipment also leaks, in imperceptible and devastating ways.

Starting in the 1930s, air conditioners commonly employed chlorofluorocarbons (CFCs) as refrigerant chemicals. CFCs—trademarked as "Freons" by DuPont—were revolutionary for refrigeration and air conditioning because they have boiling points far below room temperature. While the trademark "Freon" is said to have been inspired by the word "freeze," it also calls to mind ideas about thermal "freedom" and "free" cooling with no environmental costs. Compressed into liquid form and run through coils, CFCs and other refrigerants absorb surrounding heat when the pressure is released and they're allowed to evaporate. Refrigerant gases are typically released into the atmosphere when AC equipment leaks (refrigerant leaks are the most common mechanical problem in the AC industry) and when equipment is improperly disposed of. CFCs were the most common type of refrigerant until the 1980s, when atmospheric scientists showed that they were depleting the earth's ozone layer. In addition to their effect on the ozone layer, which was well publicized in the 1980s, CFCs are also among the most impactful greenhouse gases. As Eric Dean Wilson explains in his investigation of the connections between Freon and climate change, "CFC-12 can trap 10,200 times as much heat as an equivalent mass of carbon dioxide."[4]

The Montreal Protocol on Substances that Deplete the Ozone Layer, which prohibited the production of CFCs

---

[4]Eric Dean Wilson, *After Cooling: On Freon, Global Warming, and the Terrible Cost of Comfort* (New York: Simon & Schuster, 2021), 4.

in 1987, did not address the larger problem posed by refrigerants as climate emissions. Hydrofluorocarbons (HFCs), which became the most commonly used refrigerants following the ban, are also "highly potent greenhouse gases" projected to account for anywhere from seven to nineteen percent of global greenhouse gas emissions by 2050. In *Drawdown: The Most Comprehensive Plan Ever Proposed to Reverse Global Warming* (2017), a slate of expert researchers and policymakers ranked the regulation of refrigerants as the most potentially impactful climate solution. Yet, Wilson writes, "It's odd to me how unfamiliar refrigerants are to so many of us despite the fact that we're surrounded by them."[5] Refrigerants like CFCs and HFCs are a powerful example of air conditioning's vicious cycles: they're designed to provide temporary cooling, but when they inevitably leak into the atmosphere they become long-term contributors to planetary warming.

## AC and Environmental Injustice

The costs and benefits of mechanical "weathermaking" are distributed in deeply inequitable ways. Whether they take the form of air pollution, climate change, or waste heat that contributes to higher urban temperatures, the

---

[5]Wilson, 9, 11.

hidden externalities of air conditioning disproportionately harm vulnerable populations worldwide. People of color and the global poor are less likely to have access to either mechanical cooling or adequate medical care for chronic illnesses associated with air pollution. Women are more prone to the reproductive health effects of air pollution and extreme heat. Indigenous people and the global poor are more likely to be reliant on relationships with local food and water sources disrupted by energy extraction, drought, and air pollution. Children, the elderly, and disabled people are also at increased risk for the health effects of extreme weather and airborne toxins. Across the planet, AC redistributes life chances not only by making some spaces more comfortable and atmospherically stable for relatively privileged people, but also by contributing to environmental disruptions that most acutely affect socially marginalized people, as well as the entire spectrum of nonhuman lives.

The recent surge in wildfires across the Western United States illustrates how the consequences of AC usage can compound in ways that disproportionately harm vulnerable populations. Both climate emissions and faulty AC equipment contribute to wildfire risk, and the resulting hazards—including air pollution—are disproportionately borne by frontline workers such as the more than 1,000 incarcerated firefighters "paid $2 to $5 a day in camp and an additional

$1 to $2 an hour when they're on a fire line" in California.[6] People with respiratory conditions and other preexisting conditions are also especially vulnerable to wildfire smoke. When utility companies impose targeted, pre-emptive power outages to reduce the likelihood of wildfires ignited by electrical equipment during high-risk weather conditions, they reduce the danger of wildfire by endangering people with chronic health conditions who depend on electrically powered medical equipment.

Economic inequality also affects people's ability to transition to more efficient, less environmentally destructive AC technologies. For example, the former president of Pakistan's Heating, Ventilation, AC, and Refrigeration Society explains that after the Montreal Accord, "China used Pakistan as its dumping ground by selling us soon-to-be discarded and obsolete ACs at dirt-cheap rates. Many amongst us, who would otherwise not have afforded them, were able to buy them."[7] Another recent study indicates that in Brazil, most of the sixty percent of households that have air conditioning employ HFC refrigerants, and there is considerable room

---

[6]Jaime Low, "What does California Owe its Incarcerated Firefighters?" *The Atlantic* (July 27, 2021), https://www.theatlantic.com/politics/archive/2021/07/california-inmate-firefighters/619567/.

[7]Khurram Malik, qtd in Zofeen Ebrahim, "How Will Pakistan Stay Cool While Keeping Emissions in Check," *The Third Pole* (Mar 11, 2022), https://www.thethirdpole.net/en/climate/pakistan-cooling-action-plan/.

for increased AC energy efficiency.[8] These less efficient AC units emit pollutants that harm local communities, as well as carbon emissions that exacerbate the climate crisis. These two levels of damage—to vulnerable local communities and to the planetary climate—show that climate change can only be effectively addressed if we prioritize the needs and well-being of communities and ecologies that are most vulnerable to environmental harm.

## Outdoor Weathermaking

Although it is most commonly associated with indoor AC, the idea of "artificial weather" has also extended into the outdoors. Experiments with controlling outdoor weather include the British Royal Air Force's Fog Investigation and Dispersal Operation in 1944, the Japanese scientist Nakaya Ukichiro's research on artificial snow and rain, and Cold War cloud seeding projects such as a secret program (Operation Sober Popeye) that attempted to extend the monsoon season over North Vietnamese supply lines during the Vietnam War.[9] While many of these efforts to control the weather

---

[8]Virginie Letschert, et al., "The Manufacturer Economics and National Benefits for Cooling Efficiency for Air Conditioners in Brazil," *ECEEE 2019 Summer Study* (2019).

[9]Yuriko Furuhata, *Climatic Media: Transpacific Experiments in Atmospheric Control* (Durham: Duke University Press, 2022), 32-35.

have been led by military scientists, there is a long history of military innovations being adopted for everyday civilian applications.

Cloud seeding—the practice of dispersing chemicals such as silver iodide, potassium iodide, or dry ice into the air in an effort to increase or control precipitation—is now employed all over the world in efforts to counteract the effects of drought. For example, Western United States have been using cloud seeding for decades to increase snowpack for water supplies, hydroelectric power, and ski resorts. In recent years, Idaho, Utah, Colorado, Wyoming, and California have expanded cloud seeding projects in response to the ongoing drought. In August 2022, authorities in China (whose weather modification plans include creating a "sky river" to redirect water to more arid, northern regions[10]) decided to deploy cloud seeding in an effort to alleviate the effects of a record-setting drought on water supplies and agricultural output in Sichuan. Like air conditioning, however, cloud seeding doesn't aim to create desired weather conditions, but to redistribute them. It is at best a temporary measure against climate catastrophes; at worst, it could be used to redirect precipitation towards more privileged and powerful areas.

---

[10]James Dinneen, "Can Cloud Seeding Help Quench the Thirst of the U.S. West?" )Yale Environment 360 (Mar 3, 2022), https://e360.yale.edu/features/can-cloud-seeding-help-quench-the-thirst-of-the-u.s.-west.

Researchers have long questioned whether cloud seeding works at all, noting the lack of clear statistical evidence of increased precipitation. While a 2020 University of Colorado study established that cloud seeding does work ("at least when seeding for snow"), its lead scientist notes that "We still don't have a very great understanding of how much water we can produce."[11] Critics have also pointed out that the most common chemical used in cloud seeding, silver iodide, is known to be toxic, and that such chemicals could have harmful ecological effects as they bioaccumulate in seawater and in diverse species. China's use of small rockets for seeding clouds also carries risks: in a widely circulated 2022 video, a rocket falling from the sky nearly misses a group of pedestrians in Sichuan. As an effort to alleviate drought, cloud seeding exemplifies another vicious cycle of climate control—an effort to create artificial weather (along with new sources of environmental risk and uncertainty) outdoors in order to counteract the effects of droughts caused, in part, by the externalities of artificial indoor weather.

# Social Weathermaking

Air conditioning's influence isn't limited to the actual weather—it also has profound effects on political and social

---

[11]Ibid.

climates. Beginning in the 1950s, it enabled a mass migration and economic boom in the US South and Southwest, where it made homes and workplaces livable throughout the summer. In the Sun Belt—a loosely-defined region that spans the Southern US from Florida to Southern California—space and jobs were readily available, and homes were relatively cheap. Air conditioning enabled settlers to move onto lands that had been forcibly taken from the Navajo, Hopi, and other Indigenous people in the region, and to benefit from jobs and cheap energy made available by mining and coal plants established on Indigenous lands. Throughout the Sun Belt, air conditioning laid the groundwork for rapid urbanization and population growth: in the South, for example, the percentage of air-conditioned homes in the region grew from around ten percent in 1955 to fifty percent by the end of the 1960s.[12]

In his influential book, *Power Shift: The Rise of the Southern Rim and Its Challenge to the Eastern Establishment* (1975), Kirkpatrick Sale warned that the Sun Belt states—with an economy based in agribusiness, defense, technology, oil and natural gas extraction, construction and real estate, and tourism—were growing into a stronghold of political conservatism. Expanding upon Sale's argument, the science

---

[12]Raymond Arsenault, "The End of the Long, Hot Summer: The Air Conditioner and Southern Culture," *Journal of Southern History* 50:4 (1984), 610.

writer Steven Johnson has suggested that, by facilitating the migration of conservative retirees to the Sun Belt, air conditioning played a pivotal role in electing Ronald Reagan president in 1980. The conservative political climate enabled by air conditioning in the Sun Belt adds a social dimension to AC's vicious cycles, contributing to intensified inequality and environmental depredation, weakened environmental regulations, and the growth of increasingly fortified suburban enclaves.

But if air conditioning strengthened the economic and political influence of the South, the historian Raymond Arsenault argues that it also homogenized distinctive aspects of Southern culture. "Air-conditioning has changed the southern way of life," he writes. In addition to architecture and sleeping habits, Arsenault suggests that air conditioning has also eroded distinctive regional traditions such as "cultural isolation, agrarianism, poverty, romanticism, historical consciousness, and orientation toward non-technological folk culture, a preoccupation with kinship, neighborliness, a strong sense of place, and a relatively slow pace of life."[13] While it's debatable whether some of these characteristics are traditions or just stereotypes, Arsenault's idea that AC flattens cultural distinctiveness and "sense of place" has implications for any region where climate control is widely adopted. While marketers present AC as a source

---

[13]Ibid., 616.

of comfort, stability, and happiness, it is worth pausing to ask what a society loses—perhaps without even noticing the loss—when it takes up air conditioning.

In addition to sense of place, air conditioning also reshapes our sense of time. Thanks to the thermostat, AC can be available continuously, maintaining uninterrupted thermal comfort 24/7. Inside an air-conditioned bubble, people are cut off from natural markers of time such as the change of seasons, or diurnal changes in temperature and light. Thanks to AC, one can count on the fact that (assuming the power doesn't go out) the air will feel and smell more or less the same from moment to moment. And thanks to thermostats like Nest, that apply machine learning, you don't even have to adjust a dial. Artificial weathermaking cuts us off from the seasonal patterns that, throughout human history, have provided reference points and inspiration for cultural traditions and community gatherings.

In his "Theses on the Philosophy of History," the philosopher Walter Benjamin explains that the idea of "homogeneous, empty time"—where moments in time become empty containers for human activity and "progress"— is an essential component of capitalism. Air conditioning generates this sense of empty time, not only as a concept but as something immersive, continuously felt and inhaled from the air around us. This homogenization of time, however, cannot occur without depleting energy resources and atmospheric stability that reach far across the globe, and far into the future. The constancy of time cut off from the weather and the sun, in a particular place and time, comes at the cost of unpredictable, catastrophic weather effects in other places and far into the planetary future.

The idea of "weathermaking" puts the air conditioner in a godlike position—as Amiri Baraka said (in another context, which we'll return to later), "God has been replaced . . . by respectability and air conditioning." But, despite its considerable and comforting powers, this godlike technology demands immense and continuous sacrifices—not only vast supplies of fuel and quiet transformations of architecture and culture, but the sacrifice of possible futures and alternative ways of living in the present, and the continuous curtailment of human and nonhuman lives. Unless AC technologies can be made much more sustainable and deployed with greater equity and care, the outdoor effects of weathermaking will continue to contribute to premature deaths among the world's most vulnerable populations.

# 4 COLD STORAGE

**FIG. 5** Mika Rottenberg, *AC Trio,* 2015. Mixed media. Dimensions variable. Courtesy of the artist and Andrea Rosen Gallery, New York. Photograph by Zachary Balber. Courtesy of The Bass, Miami Beach.

Although it has come to dominate public thinking about air conditioning, human comfort was not the initial goal of engineers like Willis Haviland Carrier and Stuart Cramer. Carrier built the first electrical air conditioning unit in 1902 to control fluctuations in humidity at a Brooklyn printing company: muggy weather had been disrupting the multicolor printing process by making the size of paper vary slightly as different colors were printed on the same sheet. Cramer, who coined the term *air conditioning* in 1906, was a textile mill engineer concerned with controlling humidification in cloth factories. He modeled the term on *yarn conditioning*, which referred to the use of humid air to make textile fibers "more elastic and tough, easier to card, spin, and weave." Before the spread of air conditioning oriented towards human comfort, the industry's main concern was "process air conditioning"— controlling atmospheric conditions for the standardized manufacture of goods such as pasta, soap, fabric, chocolate, ammunition, and pharmaceuticals.[1]

Today, in addition to manufacturing processes, air conditioning helps maintain the stability of cultural objects such as oil paintings, archival documents, and digital information stored in data centers. As media studies scholar Nicole Starosielski explains, precise temperature control is essential to the production, preservation, and operation of all media, from print and analog tapes to digital data.

---

[1]Cooper, 18-19, 29-50.

Vast amounts of energy are expended to maintain optimal temperatures in museums, libraries, archives, and data centers—not for human comfort, but to maintain the material stability of objects like canvas, paint, paper, parchment, ink, and microchips. Modern spaces for exhibiting art, preserving culture, and storing data have had a powerful role in shaping public ideas about what it means to be human. Yet these spaces would not be possible without the quiet and constant presence of air conditioning. Air conditioning's invisible and indispensable role in cultural preservation challenges us reckon with the ways in which the very infrastructure of "culture" as we know it has contributed—and continues to contribute—to climate change and other forms of environmental injustice.

## Aesthetic Air

"The museum," writes Peter Sloterdijk, "can be described as a general isolator for objects: whatever there is to see or experience in it appears as an insulated artifact whose presence seeks interaction with a specialized form of aesthetic attention"[2] Modern art museums seek to isolate art not only from direct human contact, but from the chemical effects of temperature, humidity, and airborne particles. Fernando

---

[2]Sloterdijk, *Foams*, 314.

Domínguez Rubio, a communication studies professor who spent years observing and thinking about the Conservation Department at New York's Museum of Modern Art, explains:

> If you want to think about the duration, fragility, or sturdiness of artworks you need to think ecologically. For example, at 35°C and 80% RH [relative humidity], a paper-based artwork is a fleeting object that can last about 3 years. At 20°C and 50% RH, the same artwork becomes a durable object that can last for about 100 years. And at 10°C and 40% HR, it can last 1,200 years. Museums and storages are essentially huge fridges that have to be kept at constant hygrothermal conditions 24/7.[3]

Reframing museums and other institutions of cultural preservation as "huge fridges" shifts our attention from conventional modes of visual and aural engagement to something that museum visitors are generally not expected to think about at all: the invisible air and all the care and energy that goes into maintaining it.

HVAC infrastructures are essential to preventing oil paint, paper, marble, and other materials from deteriorating

---

[3]Shiv Issar and Fernando Domínguez Rubio, "Q & A With Fernando Domínguez Rubio," *SKAT: Science, Knowledge, and Technology Section of the American Sociological Association* (Feb 7, 2021), https://asaskat.com/2021/02/07/q-a-with-fernando-dominguez-rubio-author-of-still-life-ecologies-of-the-modern-imagination-at-the-art-museum/.

(or, in the case of nitrate film, exploding). Temperature affects the rates of chemical reactions, and extremes or fluctuations in humidity can physically damage materials; airborne particulates, including dust and exhalations emitted by museum visitors, can also lead to chemical or biological deterioration. The fantasy of the art object as an eternal work that stands apart from its audience makes unsustainable demands on HVAC technologies. "Producing eternity is extraordinarily expensive," explains Domínguez Rubio, because it requires the continuous maintenance of a temperature of twenty degrees Celsius and RH of fifty percent in exhibition spaces. (Based on extensive research dating back to the 1930s, this "20/50 rule" is a compromise between the thermal needs of humans and artworks.) Noting that the Metropolitan Museum of Art consumes as much electricity as 10,000 households and that the Victoria and Albert Museum estimated that it generated 7,000 tons of $CO_2$ in 2004, Domínguez Rubio asserts that "few objects are as unsustainable as the modern art object."[4] Air conditioning is also essential to offsite storage and data facilities that are seldom seen by museum visitors: "remote and low-cost warehouse[s]" where both material objects and

---

[4]Fernando Domínguez Rubio, *Still Life: Ecologies of the Modern Imagination at the Art* Museum (Chicago: University of Chicago Press, 2020), 164, 232, 256-7. Importantly, the energy consumption figures from MOMA and the Victoria and Albert Museum do not include the massive energy consumption demanded by the museums' storage facilities.

quickly growing digital collections are held in cold storage.[5] Among other things, these insights about the unsustainable atmospheres of modern art museums help to contextualize the interventions of Extinction Rebellion activists who have been gluing themselves to paintings in museums such as the National Gallery, the Prado Museum, and the Mauritshuis Museum in an effort to draw attention to the climate crisis.

At the same time that it stabilizes art objects, the carefully curated "aesthetic air" of art galleries and museums desensitizes viewers to their own bodies. In order to create conditions for a distanced, visual (or, in some cases, acoustic) experience of artworks, the deodorized and climate-controlled atmosphere of the "white cube" gallery dampens embodied senses like smell, touch, and thermoception. If, under such conditions, the artwork seems separate from the viewer, the viewer also feels separate from their environment. Paradoxically, aesthetic experience is supposed to take place in what Domínguez Rubio calls an "anaesthetic void"—an atmosphere sharply demarcated from the (often, urban) air around the museum.[6] The modern art museum imagines aesthetic experience as perception and reflection occurring in a completely neutral environment.

---

[5]Samir Bhowmik, "Thermocultures of Memory," *Culture Machine* 17 (2019), https://culturemachine.net/vol-17-thermal-objects/thermocultures-of-memory/.

[6]Domínguez Rubio, 159, 238.

In their influential *Air Conditioning Show* (1966-67), the Art & Language group (Terry Atkinson and Michael Baldwin) conceptualized an exhibition that would put the gallery's air-conditioned atmosphere itself on display. The show would consist of a single air conditioner emitting air that approximated the temperature outside. When the exhibition was installed in 1972 at the School of Visual Arts in New York, the accompanying text insisted that the room and the atmosphere should be "as bland as possible in every feature . . . visually, in its audio aspects, olfactronically." The focus of the exhibition was the "take-it-for-granted quality of the temperature," which would be maintained "by technically modifying the air-conditioning equipment to make it sensitive to any increase in temperature caused by the entrance of any person(s) into the room and to further enable it to regulate itself to ensure that the temperature would remain at a pre-decided constant normal level." The idea was to explore whether it was possible to exhibit as art something as imperceptible as climate-controlled air: "would there be a difference between a 'non-exhibition' model and a model that exhibits nothing?"[7] The *Air Conditioning Show* confronted its audience with the omnipresent, sensorially bland, desensitizing atmosphere that we are accustomed to ignore—or take for granted—when viewing art. What would experiencing art feel like under different atmospheric conditions?

---

[7]Terry Atkinson, Michael Baldwin, and Joseph Kosuth, *The 'Air Conditioning' Show 1996,* poster (New York: Visual Arts Gallery, 1972).

In her 2016 installation *AC Trio* (Fig. 5), the New York-based artist Mika Rottenberg also put air conditioning on display. But where the *Air Conditioning Show* presented AC as an anaesthetizing technology that "exhibits nothing," Rottenberg features air conditioning's externalities—the condensation drips and waste heat constantly emitted in the course of maintaining neutral air indoors. *AC Trio* consists of the *outdoor* condenser sections of three window AC units (the heat-emitting component pictured on the cover of this book) mounted close together, indoors, on a gallery wall. These units run continuously, dripping condensation onto a heated pot and pan on a platform below. The condensation drops evaporate immediately upon contact, making a sizzling sound that draws attention to the heated surfaces. Meanwhile, the outdoor condensers emit waste heat into the gallery space (which is presumably neutralized by the museum's own HVAC system). A potted houseplant sits next to these pots and pans, and another rests on one of the AC units above. By placing pots, pans, and indoor plants in the same gallery space as outdoor AC condensers, Rottenberg unsettles one of the foundational architectural features of air conditioning: the separation between indoors and outdoors. Do the hot surfaces represent city streets overheated by the urban heat island effect? Do the potted plants represent the earth's flora, now unable to escape the heat and drought brought on, in part, by indoor air conditioning? A gallery attendant at the Bass Museum in Miami (where *AC Trio* is housed, and where I visited on a hot Spring day) told me that the houseplants have to be manually watered and moved to a window every few days—they can't live without some of the sunlight screened out by museum galleries in the interest of preserving more conventional artworks. If air conditioning typically symbolizes luxury and comfort, the ominous sound of condensation evaporating in such close proximity to living plants is a constant reminder of drought. While *AC Trio's* closely packed window units invoke the context of urban apartments, the work also has site-specific elements that address the ecological consequences of HVAC infrastructures in museums, and in the tropics.

## Archival Air

Air conditioning has become an essential infrastructure for preserving cultural heritage. In addition to artworks held by museums, HVAC systems provide both thermal and chemical stability for diverse archival materials including vellum manuscripts, photographs, microfilms, pamphlets, and books. *Cold Storage*, a documentary about the Harvard Depository (HD), describes the building's architecture as "a faceless emplacement of corrugated concrete flanked by ramparts of serpentine air ducts."[8] A vast offsite storage

---

[8]*Cold Storage*, dr. Cristoforo Magliozzi (metaLAB(at)Harvard, 2015), https://vimeo.com/116603551.

facility for Harvard University's libraries, HD stores most of its holdings—over nine million items—at temperatures averaging fifty degrees Fahrenheit and thirty-five percent relative humidity. *Cold Storage* investigates libraries' growing dependence on such offsite storage facilities (other prominent examples include Yale's 7.5 million item Shelving Facility and the University of California's Regional Library Facilities), where extending the material life of information and its fragile media requires continuous and extensive climate control.

In response to the ongoing climate crisis, museum and library administrators have been revisiting long-held HVAC standards. In 1999, the *ASHRAE Handbook* introduced a new chapter of guidelines for museums, libraries, and archives. These included a table of new "Temperature and Relative Humidity Specifications for Museum, Gallery, Library, and Archival Collections" that provided information about different options for temperature and humidity fluctuations, along with potential risks associated with each option. Premised on the recognition that "*Set Points can Vary from the Standard,*" the table marked a major shift from stipulating ideal conditions for conservation to a more realistic risk management approach that allows for variance in different institutions' climate control capabilities and strategies.[9]

---

[9]Stefan Michalski, "The Ideal Climate, Risk Management, the ASHRAE Chapter, Proofed Fluctuations, and Toward a Full Risk Analysis Model,"

In 2011, Nicholas Serota, director of the Tate Galleries in London, similarly urged museum administrators to rethink environmentally unsustainable conservation practices, calling for "imaginative new solutions to resolve the dichotomy between long-term collections care and expensive environmental conditions."[10] Museums, libraries, and archives have been experimenting with mild fluctuations in temperature and humidity, as well as using environmentally responsive building design—for example, strategically placed functional windows, less thermally conductive building materials, and thick walls—to minimize dependence on mechanical air conditioning. While these adaptations are helping to reduce the climate impacts of cultural institutions, they don't fully resolve the thorny questions raised by the tension between cultural heritage preservation and climate emissions.

The extraordinary amounts of energy and emissions required to maintain museum and library holdings are not just a sustainability issue. They're a climate justice issue that can't be disentangled from the Eurocentric, patriarchal,

---

contribution to Experts' Roundtable on Sustainable Climate Management Urging museum administrators to adapt more sustainable approaches to climate control strategies (Tenerife, Spain: Getty Conservation Institute, 2007).

[10]Vanessa Thorpe, "Tate Director Sir Nicholas Serota . . .," *The Guardian* (Nov 12, 2011), https://www.theguardian.com/artanddesign/2011/nov/13/nicholas-serota-art-gallery-heating.

racist, and nationalist values that have set the terms for the collection and exhibition of artworks and archival materials. Reflecting on how archives reiterate violence against enslaved Black women, Saidiya Hartman writes: "The archive is, in this case, a death sentence, a tomb, a display of the violated body, an inventory of property, a medical treatise on gonorrhea, a few lines about a whore's life, an asterisk in the grand narrative of history."[11] Race, gender, class, ableism, and colonialism have shaped not only who could produce material records and objects regarded as artworks, but whose texts and material creations have been preserved in climate-controlled facilities. Conserving exclusionary archives and collections is a continuous decision to dedicate extensive infrastructures and resources to the maintenance of ideal atmospheric conditions. The supposedly timeless records of Western civilization turn out to be remarkably fragile, their timelessness an illusion shored up by constant energy consumption. Their *cultural emissions* contribute to the climate crisis, which disproportionately affects the very groups—women, racial minorities, Indigenous people, the global poor, disabled people—whose lives and records have been marginalized by elite cultural institutions. Meanwhile, because of the extraordinary resources required to preserve cultural materials, "Many artworks from the global south face the unsavory dilemma of languishing in poorly maintained

---

[11]Saidiya Hartman, "Venus in Two Acts," *Small Axe* 12:2 (Jun 2008), 2.

storages or ending up locked in the storages of powerful institutions of the north, which then have control over how and when these artworks can circulate."[12]

# Data Emissions

In an early episode of Sam Esmail's Emmy-nominated series *Mr. Robot* (2015-19), a group of anti-corporate hackers is faced with the seemingly impossible task of expunging global consumer debt data preserved in a militarized storage facility built into a former limestone mine. The storage facility, Steel Mountain, is a fictionalized version of Iron Mountain Incorporated, which specializes in storing everything from fine art collections to corporate data. Unable to gain direct access to financial data stored on backup tapes, the show's protagonist devises a way to destroy the tapes by hacking into Steel Mountain's HVAC system and turning up the thermostat in the storage facility. While this hack may appear to be far-fetched, cybersecurity experts have confirmed that hackers "can quickly access cooling units for [a] data center and overheat the data units."[13] This thermal vulnerability is a

---

[12]Domínguez Rubio, 180.
[13]"Data Centers Facing the Risk of Cyberattacks," *Cyble Blog* (Jan 27, 2022), https://blog.cyble.com/2022/01/27/data-centers-facing-risk-of-cyber-attacks/.

consequence of the immense heat emitted by data centers, as well as the sensitivity of data processing equipment.

While we tend to think of online data as disembodied information floating freely in "the cloud," the internet in fact runs through a vast network of data centers that store, process, and distribute information. It turns out that "the cloud" depends on physical servers, immense energy consumption, and continuously running HVAC infrastructures. The American Society of Heating, Refrigerating, and Air Conditioning Engineers (ASHRAE) recommends that data center temperatures be maintained at no more than twenty-seven degrees Celsius (eighty-one point six degrees Fahrenheit), with relative humidity between fifty and seventy percent.[14] Worldwide, over seven million data centers "use about 30 billion watts of electricity, roughly equivalent to the output of 30 nuclear power plants."[15] Data Centers are projected to account for fourteen percent of global carbon emissions by the year 2040. According to Greenpeace, "if all the data centers constituted a country of their own . . . it would be the fifth most power-hungry country in the world."[16] Online data is an especially sensitive thermal

[14]Chris Walker, "Data Centers are Facing a Climate Crisis," *Wired* (Aug 1 2022), https://www.wired.com/story/data-centers-climate-change/.
[15]James Glanz, "Power, Pollution, and the Internet," *New York Times* (Sept 22, 2012), https://www.nytimes.com/2012/09/23/technology/data-centers -waste-vast-amounts-of-energy-belying-industry-image.html.
[16]Tung-Hui Hu, *A Prehistory of the Cloud* (Cambridge: MIT Press, 2015), 79.

medium because data processing equipment is unusually vulnerable to heat, and also generates a lot of heat. Because servers generate so much heat, data centers are reliant on extensive air conditioning to prevent overheating. According to one study, cooling IT equipment can account for "over 40% of total energy consumption" at data centers.[17]

In his account of a visit to the DC11 data center in the Washington, DC region, the media scholar Jeffrey Moro describes a highly securitized space—"a bunker through which only data and air may flow."[18] The metaphor of a "bunker" is apt: many data centers are located in repurposed Cold War bunkers.[19] DC11's computer rooms run air conditioners whose cooling capacity far exceeds that of home AC units. To optimize efficiency, this data center is architecturally designed to deploy "cold aisle containment," exposing only thermally sensitive machine parts to cooled air while releasing warmer air elsewhere in the building, and outside. Moro remarks that "The life cycle of a data center is one of constant negative feedback, as the targeted application of cool air delays but can never resolve servers'

---

[17]X. Zhang et al., "Cooling Energy Consumption Investigation of Data Center IT Room with Vertical Placed Server," *Energy Procedia* 105 (May 2017), 2047.
[18]Jeffrey Moro, "Air-Conditioning the Internet: Data Center Securitization as Atmospheric Media," *Media Fields Journal* (2021), http://mediafieldsjournal.org/air-conditioning-the-internet/.
[19]Tung-Hui Hu, *A Prehistory of the Cloud* (Cambridge: MIT Press, 2015), 81.

own tendencies to overheat."[20] Data storage companies are already deeply concerned about the threat that climate change poses to their operations, even as their dependence on air conditioning makes considerable contributions to global greenhouse emissions. The HVAC hack staged in *Mr. Robot* only speeds up the overheating process that is continually deferred by the data center's intricate and energy-intensive cooling infrastructures which effectively displace heat from processing equipment into the Earth's atmosphere.

Like data centers, cryptocurrency "mining" generates value from carefully air-conditioned data processing equipment. Because crypto mining translates processing speed and duration into value, it can easily cause computer processing units to become overheated. Instead of allowing processing equipment to initiate "thermal throttling" (a safeguard that slows down performance in order to keep temperatures at safe operating levels), crypto miners typically use air conditioning to keep equipment cool and running at full speed. In August 2022, the White House reported that estimates of global electricity usage for cryptocurrency were "between 120 and 240 billion kilowatt-hours per year, a range that exceeds the total annual electricity of . . . Argentina or Australia," and that crypto-assets are consuming between 0.9 and 1.7 percent of the US's electricity usage. Although some

---

[20]Moro, "Air-Conditioning."

crypto-miners and data centers have been developing more sustainable infrastructures (for example, situating facilities in cold regions like Scandinavia, or using waste heat from data processing for home heating), these industries broadly exemplify the immense energy expenditures required to keep data cool.

As the media scholar Yuriko Furuhata shows in her study of US and Japanese experiments with climate control technologies, air conditioning has also been indispensable to computational methods for predicting the future. Forecasting things like weather patterns or stock market trends depends on immense amounts of data manipulated by advanced computers. The servers and computers all depend on environments that are carefully controlled for temperature and humidity. "Temperate indoor climates with their air-conditioned atmospheres are the hidden infrastructural supports of future forecasting," writes Furuhata.[21] As technologies for predicting the future have advanced, they have come to require increasingly more complex air-conditioning systems. This creates another vicious cycle, where climate change (to which air conditioning makes significant contributions) makes weather and economies increasingly unpredictable, even as digital data (supported by intricate climate control systems)

---

[21]Furuhata, 49.

is being mobilized to make increasingly precise predictions concerning the weather and the stock market.

# Rethinking Culture's Climates

Whether they take the form of oil paintings, parchment manuscripts, archived data, or cryptocurrency, the products of our culture require air conditioning to extend their longevity. Meanwhile, the wellbeing of human and nonhuman life on Earth demands massive reductions in energy consumption. Instead of holding up works of art and literature as "timeless" achievements to be preserved from their surroundings, perhaps we should learn to value works that embrace ephemerality (Olafur Eliasson's *Ice Watch* [2014] and some components of Kara Walker's *A Subtlety, or The Marvelous Sugar Baby* [2014], for example, were intended to melt while on display), or works (like many of the artworks, novels, and films discussed in this book) that directly engage with the climates in which they're created, stored, and consumed. In response to public outcry concerning their considerable impacts on energy consumption and climate justice, even data centers and cryptocurrency influencers have begun experimenting with more energy-efficient designs and renewable energy sources. The Bitcoin Clean Energy Initiative, for example, suggests that cryptocurrency mining can be mobilized to "accelerate the global energy transition to renewables" by providing an outlet for problems of "intermittent power supply and grid congestion" that have slowed the transition to solar and wind.[22] While they may marginally contribute to reducing these industries' greenhouse gas emissions, however, greenwashing initiatives will only be adopted haphazardly, and only insofar as they are profitable.

A more promising approach to redressing the climatic impacts of cultural conservation can be seen in the repurposing of some cultural institutions as public cooling centers during recent weather events. Jess Turtle, co-founder of the Museum of Homelessness, explains that "museums could play a vital role in mitigating the worst effects of the heat for people who are street homeless or stuck in inadequate temporary accommodation" because they are generally equipped with excellent insulation, environmental controls, and access to water and toilets. Several cities in Europe and the UK opened cultural institutions to the public during the 2022 heat wave. London's "Cool Spaces" app listed the Tate Britain museum and numerous libraries among the city's cooling locations; Bordeaux and Grenoble granted free access to public museums during the crisis, and Brussels made museums free for elderly visitors and their companions. In Louisiana, the Helis Foundation funded "Art & AC," a program that offered the public free admission to air-conditioned museums on days

---

[22] "Bitcoin Clean Energy Initiative Memorandum" (April 2021), https://bitcoin.energy/files/BCEI_White_Paper.pdf

when Fox News forecasted temperatures above ninety-five degrees Fahrenheit; though well-intentioned, this privately-funded program is shot through with contradictions: the funds can be traced back to the oil magnate William Helis, and Fox News has provided one the most visible media outlets for climate deniers. Several museums in San Francisco waived admission fees when smoke from the Camp Fire blanketed the Bay Area in 2018. Public libraries are regularly designated as cooling centers during heat waves—in 2021, "7,600 people sought shelter in public libraries over four scorching days" in Multnomah County, Oregon.[23] However, inadequate funding has left these buildings in need of significant updates and repairs—especially in lower-income areas where cooling centers play a critical role in mitigating extreme heat. While these initiatives have serious limitations (Who has access to apps? What about people with chronic conditions who are under 65? Why not open museums permanently to communities exposed to air pollution on a daily basis, as the writer and environmental justice advocate Nia McAllister suggests in "The Case for Arts Institutions as Sites of Refuge from Environmental Injustice"?), they are an important first step towards recognizing that cultural institutions should not be accountable only to the objects they preserve, but to

---

[23]Sarah Sax, "Why Libraries are Essential to Climate Justice," *Yes!* (Sep 30, 2021), https://www.yesmagazine.org/environment/2021/09/30/libraries-climate-justice.

the climates and communities around them. Cooling centers remind us that the exclusionary nature of climate-controlled bubbles is not inherent to AC technology, and that air conditioning can be mobilized in more equitable and public-minded ways.

# 5 THE RACIALIZATION OF COOLING

**FIG. 6** *Do the Right Thing* (dir. Spike Lee, 1989).

Invented during the period some historians refer to as the nadir of American race relations, air conditioning quickly became recognized as a technology of racial differentiation, or what Eric Dean Wilson calls "the racialization of cooling."[1]

---

[1] Wilson, 32.

Many of the key sites of public access to air conditioning in the United States, including downtown movie palaces, high-end hotel lobbies, restaurants, department stores, and suburban shopping malls, have been spaces of legal and de facto segregation. These spaces of thermal segregation are part of a longer history of efforts to leverage temperature and weather as a tool of racism, which stretches from the sweltering hold of the slave ship and the overheated conditions of plantation labor to the streets of the "urban heat island," and to homes and prison cells with limited access to ventilation and air conditioning. How has air conditioning been used to intensify and legitimize racial inequalities in the US? To answer this question, we have to look not only at how coolness has been represented and marketed as a racial characteristic, but also at stories that focus on what the *lack* of air conditioning feels like in racialized communities.

# Climate Determinism

In the early twentieth century, manufacturers seeking to persuade Americans about the desirability of an air-conditioned future drew on misguided ideas about climate determinism popularized by academic publications such as Ellsworth Huntington's *Civilization and Climate* (1915), S. Colum GilFillan's "The Coldward Course of Progress" (1920), and Sydney Markham's *Climate and the Energy of*

*Nations* (1942). The early twentieth-century public health professor Charles-Edward Amory Winslow repeatedly invoked the Black Hole of Calcutta—a hot, unventilated dungeon in which 123 English prisoners of war were said to have died of suffocation or heat exhaustion in 1756—in his influential writings on ventilation and air-conditioning.[2] This example—which had been widely cited in nineteenth-century ventilation manuals—framed heat and poor ventilation in racialized terms, while suggesting that these are especially perilous for white, English men.

Climate determinists built on nineteenth-century ideas about climate and race that were used to justify slavery and colonialism, and deployed antiquated racial pseudoscience when arguing that hotter climates resulted in indolence and decreased vitality. According to this logic, the energy necessary to sustain human civilization and "progress" was supposedly available only in cold or temperate regions like Western Europe and the northern United States. Huntington associated cold climates with greater "mental and physical energy," identifying the "mental optimum" temperature as thirty-eight degrees Fahrenheit and the "physical optimum" as "60°F or possibly 65°F." "In general," he speculated, "the lower types of life, or the lower forms of activity, seem to reach their optima at higher temperatures than do the more advanced types and the more lofty functions such as

---

[2] Ackermann, 41.

mentality. . . . The law of optimum temperature apparently controls the phenomena of life from the lowest activities of protoplasm to the highest activities of the human intellect."[3]

These racist and Eurocentric ideas about the effect of climate on human capacity are on full display in "Boomtown—2000 B.C.," a full-page, color advertisement for the Carrier Corporation printed in the *Saturday Evening Post* (Carrier Corporation did not agree to having this image reproduced here; it can be found in *Saturday Evening Post* 222: 7 [Aug. 13, 1949], 15). The ad, which shows two Bedouins in a desolate desert landscape, reads:

> Once a thriving city—now deserted ruins! These seven words sum up the history of many a torrid zone metropolis, laid low by the combination of heat itself and conquest by stronger races from temperate climes. In ancient days man just had to accept the penalties of equatorial climate.[4]

The ad suggests that a hot, desert climate results in both racial and civilizational "penalties," leaving inhabitants vulnerable both to hot weather and to other races whose supposed superiority comes from inhabiting cooler climates. To drive the point home, the ad contrasts the Bedouin slouching on a

---

[3]Ellsworth Huntington, *Civilization and Climate* (New Haven: Yale University Press, 1915), 105, 129, 109-10.
[4]Qtd in Ackermann, 8.

camel with images of industrious whiteness: a white woman hanging laundry and a white man working at a factory, both standing upright.

However, not all white people inhabited the same climate. Whereas (at least until recently) most of Europe has a temperate climate, many Americans endure uncomfortable levels of heat and humidity. If cool temperatures had been important to the supposed "vitality" and "civilization" of white people, then Southern and Southwestern summers in the US threatened to sap the energy and potential that European immigrants had inherited from colder climates.[5] As a cautionary tale, "Boomtown" invokes fears of white racial decline in the face of America's hot climates. But the advertisement also presents a technological solution: air conditioning promised to compensate for the US's intemperate heat, cooling indoor spaces so that Americans (some of them, at least) could avoid the supposed mental and physical drawbacks associated with high temperatures.

Air conditioning offered its disproportionately white consumers an atmosphere of comfort that could be taken for granted—not just cool air, but a domestic bubble where (as we saw in Chapter 2) heteronormative families could spend time at home among clean white linens, separated from the noises and smells of the city. Air regulated by a thermostat is a self-effacing privilege, a material benefit that fades into

---

[5]Sydney Markham, *Climate and the Energy of Nations* (Oxford: Oxford University Press, 1944), 192.

the background (except when the AC breaks down). Since both exposure to extreme temperatures and access to AC are racially stratified in the US, white Americans literally inhabit different atmospheres than many BIPOC (Black, Indigenous, and People of Color). Air conditioning maintains ventilation and ambient temperatures conducive to what Robin DiAngelo calls "white equilibrium"—"a cocoon of racial comfort, centrality, superiority, racial apathy, and obliviousness."[6] What do cities and climates increasingly shaped by air conditioning feel like for those outside the cocoon?

# Thermal Racism and the Urban Heat Island

While intentional and symbolic instances of racism tend to capture the most media attention, heat-related illness and death draw attention to more insidious, structural modes of racism. Because heat exposure is determined by subtle and easily-overlooked factors like urban planning, building design, AC access, and electricity costs, its violent and racist effects are frequently overlooked. Yet heat-related deaths in the US (and worldwide) are a clear instance of the geographer Ruth Wilson Gilmore's influential definition of

---

[6]Robin DiAngelo, *White Fragility: Why It's So Hard for White People to Talk About Racism* (Boston: Beacon Press, 2018), 112.

racism: "the state-sanctioned and/or extra-legal production and exploitation of group-differentiated vulnerability to premature death."[7] As a deadly component of racism, heat vulnerability stretches from the transatlantic slave trade to twenty-first century prisons and detention facilities.

Historically, the distribution of cooling infrastructure has enhanced existing forms of racism. From the insufferable and deadly heat of the slave ship's hold to the employment of enslaved workers who spared white people from the heightened body temperatures of plantation labor, slavery subjected Black people to extreme heat. In his classic account of the Middle Passage, Olaudah Equiano describes the unbearable stench and stifling air of the unventilated hold: "The closeness of the place, and the heat of the climate, added to the number in the ship . . . almost suffocated us. . . . The air soon became unfit for respiration, from a variety of loathsome smells, and brought on a sickness among the slaves, of which many died."[8] On a later occasion, Equiano reports being called to fan his master while he slept. As with the many enslaved people who operated handheld fans and punkahs (hand operated ceiling fans) throughout the US South, Equiano's physical work heated his own body while

---

[7] Ruth Wilson Gilmore, *Golden Gulag: Prisons, Surplus, Crisis, and Opposition in Globalizing California* (Berkeley: University of California Press, 2006), 28.
[8] Olaudah Equiano, *The Interesting Narrative of Olaudah Equiano* (New York: Penguin, 2003 [1789]), 58, 63.

providing cooling and ventilation for another.[9] In the fields, too, climate determinism lent legitimacy to slaveholders who coerced enslaved people to labor in hot climates so that white people wouldn't have to. (In other contexts, similar thermal hierarchies would later be imposed on racialized laborers such as sharecroppers, Chinese railroad workers, convict laborers, and Latinx migrant farmworkers.) After emancipation, racial segregation denied Black people safe access to "public" cooling infrastructures such as refrigerated drinking fountains, beaches, swimming pools, and air-conditioned spaces in trains, restaurants, and theaters. Denying access to these means of thermal regulation was a deliberate method of racial segregation, an effort to reinforce the color line in thermal terms. On the local level of neighborhoods and individual bubbles, the uneven distribution of cooling was an effort sought to realize the climatic differences that supposedly explained humanity's racial and civilizational distinctions.

Many US cities and suburbs are susceptible to extreme heat, whether by virtue of their design or because they are located in subtropical and desert climates with the assumption that inhabitants would have access to AC. But, in practice, access to AC has been deeply uneven along

---

[9]The Punkah Project, an online resource created by Bowdoin College researcher Dana Byrd, documents the widespread and ostentations use of punkahs operated by enslaved workers in the antebellum U.S. https://research.bowdoin.edu/punka-project/abstract/.

class and racial lines. Moreover, Black and Brown urban communities are disproportionately exposed to the "urban heat island effect," or the way in which temperatures in urban spaces are increased by interacting factors including the thermal qualities of building materials, the trapping of heat by air pollution and groups of tall buildings, lower levels of evaporation as a result of drainage systems, and the concentration of artificial heat sources (including waste heat from air conditioners).

In addition to causing heat-related discomfort, health complications, and fatalities, the urban heat island can sometimes create what is called a temperature inversion, where cooler and denser air near ground level—along with smoke and other airborne pollutants—get trapped under a layer of warmer air above it. The result is a kind of unintentional bubble filled with toxic materials. Temperature inversions have been linked to higher incidences of asthma and other respiratory conditions, cardiovascular diseases, and birth abnormalities. These weather events have become increasingly common as climate change has intensified in the past few decades, and (because they prevent existing toxins from dispersing) their effects are most hazardous in communities that are already exposed to greater levels of air pollution.

Heat islands have tended to track with the economic and racial disparities that stratify American society. Through practices like racially restrictive housing covenants (which prevented suburban homes from being sold to BIPOC buyers), redlining (in which banks denied services such as loans and insurance to people living in neighborhoods deemed "high risk"), and white flight (where millions of white households moved from increasingly diverse urban neighborhoods to segregated suburbs beginning in the 1950s), many Black and Brown people have come to inhabit buildings and urban areas that are especially susceptible to the urban heat island.

A 2020 analysis of 108 US urban areas found strong correlations between historically redlined neighborhoods and uncharacteristically high temperatures. In ninety-four percent of the areas studied, the authors identified "patterns of elevated land surface temperatures in formerly redlined areas relative to their non-redlined neighborhoods by as much as 7°C [12.6°F]."[10] In 2021, another study focusing on the US Southwest found significant disparities between neighborhoods with versus without Latinx populations: on extreme heat days, heavily Latinx neighborhoods in the Inland Empire and Los Angeles were six point five degrees Fahrenheit hotter than the least Latinx areas.[11] Historically redlined neighborhoods are more likely to have inadequate tree shade and an overabundance of heat-absorbing asphalt surfaces; meanwhile, residents in these dense urban areas

[10]Jeremy Hoffman et al., "The Effects of Historical Housing Policies on Resident Exposure to Intra-Urban Heat: A Study of 108 US Urban Areas," *Climate* 8:12 (2020), 1.
[11]John Dialesandro et al., "Dimensions of Thermal Inequity," *Int. J. Environ. Res. Public Health* 18:3 (2021).

can be dissuaded from opening windows by factors such as outdoor noise, air pollution, and fear of crime. Lower incomes and less access to generational wealth also result in considerably lower levels of AC availability and usage, leaving people with little respite from extreme heat. In the first (1960) Census to ask about air conditioning, eighteen percent of all households in the US South had AC, compared with just three point eight percent of nonwhite households in that region; nationwide, twelve point four percent of all households had AC, compared with only four percent of nonwhite households.[12]

These patterns of thermal racism explain why, in the deadly 1995 Chicago heat wave, "African Americans had the highest proportional death rates of any *ethnoracial* group." Urban abandonment, deindustrialization, inadequate infrastructure, substandard housing, poverty, violent crime, withdrawal from public spaces, and the heat island effect combined to make elderly Black residents especially vulnerable to dying alone in homes without air conditioning. According to Eric Klinenberg, a sociologist who spent years studying the catastrophe, "Heat waves receive little public attention not only because they fail to generate the massive property damage and fantastic images produced by other weather-related disasters, but also because their victims are primarily social outcasts—the elderly, the poor, and the

---

[12]Ackermann, 137.

isolated—from which we customarily turn away."[13] This public apathy towards the heat-induced deaths of socially vulnerable people is especially striking because, in the United States, "heat waves kill more people on average than all other natural disasters combined."[14]

# Thermal Punishment

Prisons and detention centers are another site where the psychological and physical violence enacted by temperature receives little public attention. In Texas, where prison temperatures often exceed one hundred degrees Fahrenheit in the summer, at least thirteen men died of heat-related causes in prisons between 2007 and 2013.[15] In a survey conducted by the Texas Prisons Community Advocates and Texas A&M University between 2018 and 2020, many inmates "reported a barrage of illnesses, including heat cramps, rashes, migraines

---

[13]Eric Klinenberg, *Heat Wave: A Social Autopsy of Disaster in Chicago* (Chicago: University of Chicago Press, 2002), 18, 17.
[14]Ashley Dawson and Aurash Khawarzad, "Hot City: New York City Will Never Be the Same Again—and it Shouldn't Be," *Verso Blog* (Aug 20, 2020), https://www.versobooks.com/blogs/4829-hot-city-new-york-city-will-never-be-the-same-again-and-it-shouldn-t-be.
[15]Cameron Langford, "Dying of Heat Stroke in Texas Prisons," *Courthouse News* (Jun 17, 2013), https://www.courthousenews.com/dying-of-heat-stroke-in-texas-prisons/.

and repeated fainting or trouble breathing."[16] Despite over a decade of lawsuits and studies arguing that the conditions of prison facilities are a form of cruel and unusual punishment, only about thirty percent of the state's one hundred prisons have air conditioning. In Florida, where the heat index easily reaches over one hundred degrees in the summer, "only about 24 percent of housing units in state-run prisons have air conditioning."[17] States like Florida, Texas, and Louisiana are among the hottest in the nation, and also among those with the highest incarceration rates. While there have been scattered court decisions that affirm air conditioning as an entitlement for incarcerated people faced with extreme and prolonged temperatures, inmates and their advocates are still fighting for livable thermal conditions as a basic human right throughout the US as planetary temperatures continue to rise. In the last few years, inmates have been transported in windowless metal van cabins on hot days, prison staff have fabricated or failed to record temperatures on extremely hot days in Arizona, and prisoners have continued suffering from

---

[16]Jolie McCullough, "It's a Living Hell: Scorching Heat in Texas Prisons Revives Air-Conditioning Debate," *Texas Tribune* (Aug 24, 2022), https://www.texastribune.org/2022/08/24/texas-prisons-air-conditioning/.

[17]Amanda Rabines, "Prison Reform Activists Demand A/C Amid 'Sweltering Heat' within Florida State Prisons," *Orlando Sentinel* (Jul 23, 2022), https://www.orlandosentinel.com/news/breaking-news/os-ne-protest-lake-eola-air-conditioning-prisons-20220723-h6ulk2rdbfbrvefwaws5ia2fmm-story.html.

a range of heat-related health conditions in inadequately air-conditioned facilities.

Deadly temperatures in prisons are another example of racism's influence on thermal disparities in the United States. According to the NAACP, African Americans are 5 times more likely to be incarcerated than white people; according to the Bureau of Justice Statistics, Native Americans are incarcerated at more than double the rate of white people. In her work on thermal media, Nicole Starosielski has traced the use of temperature as a means of punishing incarcerated Black and Brown people to the practice of sweatboxing. The sweatbox, a small wooden box used to expose a person to extreme and prolonged cold or heat, was "used by colonizers to torture Native Americans" and subsequently taken up by Southern slaveholders (and overseers) as a means of torturing enslaved people. The practice of sweatboxing continued after Emancipation, "especially in the prisons and on the chain gangs that formed on the new plantations," and primarily in punishments directed at Black people. Although sweatboxing was used to murder Black inmates on numerous occasions in US prisons, these deaths were underreported in the media and "were often attributed to an accident or 'natural causes.'"[18] Although the use of sweatboxes officially ended in 1958, prison architectures and staff have continued

---

[18]Starosielski 116, 118, 120.

leveraging extreme cold and heat to punish and demoralize inmates.

These techniques of thermal torture have spread from prisons to militarized detention centers. In 2005, former and current CIA officials described an interrogation technique known as "Cold Cell" as follows: "The prisoner is left to stand naked in a cell kept near 50 degrees. Throughout the time in the cell the prisoner is doused with cold water."[19] According to the Associated Press, a detainee named Gul Rahman died of hypothermia in a CIA black site after being stripped from the waist down and left in a cold cell. At Guantánamo Bay, air conditioners were turned up and down in Mohammed al Qahtani's unventilated cell, subjecting him to both freezing cold and temperatures over one hundred degrees Fahrenheit. His lawyer, Gitanjali Gutierrez, declared that "he was placed in rooms with very cold temperatures and to this day is sensitive to cold temperatures during attorney client meetings," and a General testified that "at times [he] suffered from hypothermia." Along with the painful restraints and positions that al Qahtani was put in, Gutierrez argued that "the temperature extremes" caused physical and psychological stresses that "rise to a level of torture under

---

[19]Brian Ross and Richard Esposito, "CIA's Harsh Interrogation Techniques Described," *ABC News* (Nov 18, 2005), https://abcnews.go.com/blotter/investigation/story?id=1322866.

international law."[20] This testimony is consistent with the International Committee of the Red Cross's leaked secret report, which charged that "Detainees frequently reported that they were held for their initial months of detention in cells which were kept extremely cold, usually at the same time as being kept forcibly naked. The actual interrogation room was often reported to be kept cold."[21]

At the US-Mexico border, migrants have testified that they were exposed to prolonged cold temperatures in intensely air-conditioned rooms, or *hieleras* ("iceboxes"). Here, border authorities use AC to extend the opposite of a warm welcome: a traumatic, demoralizing, and physically debilitating exposure to extreme cold. Perversely, border agents have been weaponizing AC—something frequently out of reach for undocumented migrants—to torture hundreds of thousands of immigrants, including children, many of whom have been displaced by climate change.

Recounting his experience of the *hielera* in *Buzzfeed News*, Irani Garcia Zacarias recalls being driven to "a huge compound—the hielera. … I exited the heat of the desert and entered a frigid cage. It felt like walking into a refrigerator."

[20]"Declaration of Gitanjali S. Gutierrez, Esq., Lawyer for Mohammed al Qahtani" (New York: Center for Constitutional Rights, n.d.), 21, https://ccrjustice.org/files/Gutierrez%20Declaration%20re%20Al%20Qahtani%20Oct%202006.pdf.
[21]Stan Cox, "Militarism, Torture . . . and Air Conditioning?" *Counterpunch* (Apr 22, 2010), https://www.counterpunch.org/2010/04/22/militarism-torture-and-air-conditioning/.

Detained with about 30 other migrants in this small, cold cell, Zacarias reports that "my hands turned purple and I could no longer feel anything. The cold, we learned from the laughing guards, was calculated, meant to punish those who had come and force us to self-deport."[22] In another testimonial, a girl recounts, "'My sister was shaking from the cold. Her lips were always blue.' . . . The officer told the children to stop crying or else they would 'turn up the air conditioning and make it colder.'"[23] Although it didn't involve air conditioners, a similar pattern of thermal cruelty motivated Texas governor Greg Abbott's decision to have well over one hundred migrants—many of them without adequate winter clothing—dropped off outside Vice President Kamala Harris's residence in Washington, DC in freezing weather on December 24, 2022.[24]

Scholars like Klinenberg and Starosielski have brought together knowledge and approaches from the sciences, social sciences, and humanities in an effort to understand the complex social dynamics of thermal inequality. In addition to the class and racial vulnerabilities made evident

---

[22]Irani Garcia Zacarias, "Opinion: I walked Through the Desert to Make it to America. Then They Took Me to the Icebox," *Buzzfeed News* (Aug 17, 2019), https://www.buzzfeednews.com/article/iranizacarias/first-desert-then-ice-box-hielara.

[23]Qtd. in Starosielski, 110-11.

[24]Noah Gray, "More Migrants Dropped Off . . .," *CNN* (Dec 26, 2022), https://www.cnn.com/2022/12/24/politics/migrants-dropped-off-vice-president-christmas-eve/index.html.

by heat wave deaths, UCLA's recently established Heat Lab, led by anthropologist Bharat Venkat, researches the distinctive ways in which heat affects disabled people, the complex intersections between heat and racism, the thermal experiences of food truck workers, and the "politics and science of the ongoing heat crisis in prisons and detention facilities."[25] Refusing to take temperature for granted as either a purely natural or an entirely objective phenomenon, the emerging academic field of "critical temperature studies" invites us to think hard about how perceptions and responses to temperature are deeply enmeshed with historical, cultural, and social experience.[26]

## Thermal Disparities in Black Literature and Culture

African American literature and film offers another rich resource for understanding everyday experiences of the synergies between racism and thermal inequality, often in the absence of air conditioning. Long before the term "urban heat island" was first coined, influential novels by Nella Larsen and Chester Himes explored the quiet yet powerful influences of urban heat and the ways it compounded existing

---

[25]"About," UCLA Heat Lab (n.d.), https://heatlab.humspace.ucla.edu/about/.
[26]Starosielski, 8.

racial inequities. Towards the beginning of *Passing* (1929), Larsen extensively describes Irene Redfield's encounter with the urban heat island effect on a summer day in Chicago. She emphasizes how construction materials like pavement, glass, and car-tracks amplify the heat:

A brilliant day, hot, with a brutal staring sun pouring down rays that were like molten rain. A day on which the very outlines of the buildings shuddered as if in protest at the heat. Quivering lines sprang up from baked pavements and wriggled along the shining car-tracks. The automobiles parked at the kerbs were a dancing blaze, and the glass of the shop-windows threw out a blinding radiance. Sharp particles of dust rose from the burning sidewalks, stinging the seared or dripping skins of wilting pedestrians. What small breeze there was seemed like the breath of a flame fanned by slow bellows.

Larsen's description shows how heat warps the senses as Irene's vision shudders, quivers, wriggles, dances, and wobbles. Heat numbs her empathy and sociability so that she doesn't respond when she sees a man collapse from heat exhaustion. Instead, Irene abruptly leaves the scene in a cab, "feeling disagreeably damp and sticky and soiled from contact with so many sweating bodies." Her impulse to distance herself from the sweaty crowd prompts Irene's decision to pass as white. Stepping off an elevator to a cool, breezy hotel rooftop, she feels like she has been "wafted upward on a magic carpet to another world, pleasant, quiet, and strangely remote from the sizzling one that she had left below."[27] By temporarily passing as white, Irene is able to transition from the debilitating heat of the streets to a space of thermal comfort.

Where *Passing* offers a psychological portrait of a character navigating temperature differentials and the color line, Himes's *The Heat's On* (1966) describes an entire neighborhood transformed by heat. Instead of rehearsing crime fiction's tendency to focus on individual "criminal" actors, the novel's first pages set the scene for a more complicated view of the environmental conditions for "criminality" by describing the stifling heat as a powerful influence on human actions:

> The heat had detained them.
>
> Even at past two in the morning, "The Valley," that flat lowland of Harlem east of Seventh Avenue, was like the frying pan of hell. Heat was coming out of the pavement, bubbling from the asphalt; and the atmospheric pressure was pushing it back to earth like the lid on a pan.
>
> Colored people were cooking in their overcrowded, overpriced tenements; cooking in the streets, in the after-hours joints, in the brothels. [...]

---

[27]Nella Larsen, *Quicksand and Passing,* ed. Deborah McDowell (New Brunswick: Rutgers University Press, 1996 [1929]), 146, 147.

An effluvium of hot stinks arose from the frying pan and hung in the hot motionless air, no higher than the rooftops—the smell of sizzling barbecue, fried hair, exhaust fumes, rotting garbage, cheap perfumes, unwashed bodies, decayed buildings, dog-rat-and-cat offal, whiskey and vomit, and all the old dried-up odors of poverty. [. . .]

It was too hot to sleep. Everyone was too evil to love. And it was too noisy to relax and dream of cool swimming holes and the shade of chinaberry trees.[28]

This tour of Harlem's hot places and "hot stinks" leaves us with a strong sense of Harlem's inescapable heat. The neighborhood becomes a hellish "frying pan" or pressure cooker, where heat rises from construction materials and is pushed back downwards by "atmospheric pressure" that is caused, in part, by urban air pollution. The heat also brings out repulsive smells. All this combines to make residents feel "too evil to love," as the weather (combined with inadequate infrastructure and the absence of air conditioning) gives rise to a shared psychological state. By the end of the scene, Himes's hardboiled detectives are detained not only by "the heat," but by its social outcome: "an outburst of petty crime." The novel represents heat as both a metaphor and a psychological disposition towards violence and theft. For

---

[28]Chester Himes, *The Heat's On* (New York: Vintage, 1988 [1966]), 23-4.

Himes, urban crime is not perpetrated by individuals acting in a vacuum, but by people immersed in hot, putrid air.

Spike Lee's iconic film, *Do the Right Thing* (1989), uses heat to convey communal feelings of exhaustion, dissatisfaction, and rage. The film's hot setting was a deliberate choice. Lee responded to the Howard Beach Incident—a deadly mob attack on three Black men in Queens that occurred in the *winter* of 1986—by making a film set on the hottest day of the year, largely in a pizzeria with a broken AC. Reminiscing about the film in 2019, Lee comments, "I want[ed] audiences to be sweating even though the motherfucking theatre is air-conditioned."[29] In a 1988 journal entry, Lee recounts a discussion with his cinematographer, Ernest Dickerson, about how important heat was to their project:

> He's fired up. He's already thinking about how to visualize the heat. He wants to see people in the theaters sweating as they watch the film. . . . Every character must comment on the heat. . . . Anytime the camera is rolling we should be thinking about the heat. I want to have sequences in *Do*

---

[29]Chris Hewitt, Spike Lee, John Turturro, Giancarlo Esposito, "Spike Lee's *Do the Right Thing*: An Oral History," *Empire* (Mar 4, 2020), https://www.empireonline.com/movies/features/empire-30-spike-lee-do-the-right-thing-oral-history/.

*the Right Thing* where we suspend the narrative and show how people are coping with the oppressive heat.[30]

Dickerson explains that he used color to communicate heat visually: "The warmer colours have the tendency to raise the heart rate."[31] In one groundbreaking shot, he placed a heat bar under the camera lens to simulate the wavy image we see when light waves are refracted by intense heat.

Although he was influenced by the idea that violent behaviors correlate with hot weather, Lee communicates the physical, biochemical, and emotional effects of extreme heat without reducing the film's Black characters to victims of climate. In *Do the Right Thing*, heat isn't just something that people suffer through or react to—it's a condition of knowledge. In addition to presenting temperature as a precondition for the police murder of Radio Raheem and the ensuing Black uprising in the film's concluding scenes, *Do the Right Thing* also uses montage to pay tribute to everyday creative acts of heat regulation—for example, people dunking their heads in cold water, lovers experimenting with ice cubes in bed, or neighborhood kids reveling in the stream of an open fire hydrant—that mitigate thermal violence. In the iconic fire-hydrant scene (Fig. 6), the community's capacity for relief and revelry in the midst of a heat wave is eventually

---

[30]Spike Lee and Lisa Jones, *Do the Right Thing: A Spike Lee Joint* (New York: Fireside, 1989), 50.
[31]Hewitt et al., "Spike Lee's *Do the Right Thing*: An Oral History."

shut down by the police, who issue threats as they turn off the flow of water. Partially shaded by an umbrella on an overheated Bed-Stuy sidewalk, one gentleman, dressed in heat-reflective white clothing, proclaims that the warming climate will melt the polar ice caps and flood the world. Heat influences the enraged crowd in the film's last scenes, bringing to mind Claudia Rankine's insight that anger can be "really a type of knowledge"[32]—in this case, a rational response to the quiet ways that urban planning and climate change target Black communities. Across films like *She's Gotta Have It* (1968), *Crooklyn* (1994), and *Red Hook Summer* (2012), Lee explores how communities get by—and even thrive—amid New York's urban heat islands. As one group of critics puts it, "Lee is the poet of American summers, and he's here to tell the story of love vs. heat."[33]

Lee is not alone in finding knowledge and poetry in urban heat. Without minimizing the serious health impacts of heat inequality or the importance of recognizing air conditioning access as a human right in extreme weather conditions, Black communities and artists have responded to heat with the culture of urban "cool," with social uprisings (for example, during the "long, hot summers" of the 1960s), and with

---

[32]Claudia Rankine, *Citizen: An American Lyric* (Minneapolis: Graywolf, 2014), 24.

[33]Forrest Wickman, Aisha Harris, and Holly Allen, "Introducing the Spike Lee Heat Index!" *Slate* (Aug 10, 2012), https://slate.com/culture/2012/08/the-spike-lee-heat-index-the-hotter-the-spike-lee-movie-the-better-it-is.html.

imaginative explorations of heat as a transformative force. We can see heat as a generative force for Black expression in the work of Amiri Baraka, who imagined the Black Arts movement as a necessary "flash of heat" that would help spark a social revolution. For Baraka, artificial cooling has taken over Christianity's role as a force of political pacification: "God has been replaced . . . by respectability and air-conditioning."[34] Elsewhere he remarked, "The slave ship grew more sophisticated and gave a few Negroes radios or air-conditioning in the hold."[35] Writing more recently about Blackness and climate change, the poet Alexis Pauline Gumbs imagines heat as a force of volatility and transubstantiation: "*Something can turn into anything if you get it hot enough.*"[36]

Something like this happens in a 2019 installation by Ima-Abasi Okon, a conceptual artist based in London and Amsterdam. The installation takes the form of a room with an unusually low modular ceiling, made of the familiar interchangeable squares used in institutional interiors. On the walls are eleven AC condenser units (the ones usually located on the outside of buildings), their fans rotating at different speeds. But instead of emitting waste heat, these AC components surprise us. As one critic writes, "the air

---

[34]Amiri Baraka, "What Does Nonviolence Mean?" in *Home: Social Essays* (New York: Akashi, 2009 [1966]), 168.

[35]Amiri Baraka, "Street Protest," in *Home: Social Essays*, 118.

[36]Alexis Pauline Gumbs, *M Archive* (Durham: Duke University Press, 2018), 96.

conditioners suddenly vary the speed and intensity of their mechanical humming . . . and music(!) comes out of them."[37] The song, slowed down into a "dragged-out melody of distorted song," is barely recognizable as music. Okon's work—its full title is *(Unbounded [sic]-Vibrational [sic] Always [sic]-on-the-Move [sic]) Praising Flesh (An _Extra aSubjective p,n,e,u,m,a-mode of Being T,o,g,e,t,h,e,r)*—puts us at the threshold of urban heat (the condenser units that increase outdoor temperatures while suffusing indoor spaces with cool comfort) being transfigured into something else. Is it possible to experience exclusion from the air-conditioned bubble—symbolized by the eleven AC condensers—as something other than dehumanization? While the title of Okon's work refuses to yield any single meaning, it hints that the music emerging from these AC units could be a call to come together out of our bubbles into a collective "breath" or "spirit" (*pneuma,* in Greek). If cooled bubbles enable their constant consumers to forget inconvenient truths like climate change, thermal inequality, and antiblack racism, heat could be a generative condition of knowledge, cultural expression, and social transformation.

---

[37]Fernando Domínguez Rubio, "The Art of Opacity," in *Ima-Abasi Okon: A Reader,* ed. PSS (London: PSS: PressforPractice, 2021), 17.

# 6 GLOBAL AC & THE GREAT UNCOOLED

**FIG. 7** poster for *Air Conditioner*, dir. Fradique. Photograph by Rui Magalhães. Courtesy of Geração 80.

Although much of this book has been focused on air conditioning in American contexts, the effects and inequities of air conditioning are increasingly experienced worldwide. As twentieth-century novelist Henry Miller implied when he referred to the US as "the air-conditioned nightmare," the entire nation can be seen as an air-conditioned bubble, albeit one shot through with thermal disparities. How do people across a broad range of international contexts experience and think about air conditioning? How do American patterns of AC use compare with its adoption in the Global South, where hot climates and the socioeconomic legacies of colonialism combine to make populations especially vulnerable to the effects of thermal inequality? On a global scale, where is the energy for AC sourced, and where are its externalities most acutely felt?

# The Global North

The founding prime minister of Singapore, Lee Kuan Yu, attributed the young nation's economic success to air conditioning, calling it "a most important invention for us, perhaps one of the signal inventions of history. It changed the nature of civilization by making development possible in the tropics."[1] Echoing the claims of earlier climate determinists,

---

[1]Qtd in Ognjen Miljanic and Joseph Pratt, *Introduction to Energy and Sustainability* (Wiley, 2022), 393.

Lee claimed that AC set the stage for rapid development and productivity by making thermal comfort available throughout the day in tropical regions. In wealthy Southeast Asian commercial centers like Singapore and Hong Kong, AC is almost universal: ninety-nine percent of condos and most homes have at least one AC unit. In Singapore, AC accounts for sixty percent of energy consumption in nonresidential buildings. AC is also widely available in Japan (it's in ninety percent of homes—the same rate as in the US). While AC has been widely adopted in these countries, many homes in Asian cities employ smaller, single-room units (which can be considerably more efficient, cooling only spaces that are being used) instead of central air.

Even among Asian countries where AC is widely available, there are significant differences in attitudes and regulations. As Hong Kong's average temperature has risen by about a quarter degree Celsius per decade between 1991 and 2020, air conditioning usage has become a habit among many residents. One journalist charges Hong Kongers with "air con abuse," noting that that many people wear "indoor jackets" year-round to keep warm amid ACs set to low temperatures, and that buildings often run the AC even during winter to maintain air circulation and an appealing shopping ambience.[2] (While articles like this one offer helpful context, Westerners'

---

[2]Anna Cummins, "Hong Kong's Air Con Abuse," *Time Out* (May 20, 2016), https://www.timeout.com/hong-kong/blog/hong-kongs-air-con-abuse-its-not-cool-052016.

complaints about Hong Kongers' AC usage and energy consumption can deflect attention from the American origins of AC, and the Western origins of a global economy founded on fossil fuel combustion.) In Taipei, by contrast, businesses are prohibited from setting their ACs lower than seventy-nine degrees Fahrenheit. While South Korea also has high rates of AC ownership (eighty-six percent, according to the International Energy Agency), many South Koreans perceive mechanical ventilation as a health threat. This uneasiness can be traced to the idea of "fan death," or deaths allegedly caused by the breeze from a mechanical fan, which began to be reported in the 1920s. In 2005, a press release from the government-funded Korea Consumer Protection Board cited "asphyxiation from electric fans and air conditioners" as a leading cause of recent summer injuries. Many South Koreans also believe that air conditioning can cause *naengbangpyeong,* or "air-conditioningitis," a condition whose reported symptoms include dryness in the nose and throat, feeling cold, dizziness, fatigue, indigestion, constipation, diarrhea, and swelling. With frequent news reports warning of the dangers of AC, South Koreans as a whole take a more conservative approach to cooling than Hong Kongers.

Whether because of more temperate climates, cultural preferences, or economic limitations, many nations have taken a moderate approach to AC adoption. Compared with the US, where sixty-six percent of households have central air and another twenty-two percent use individual air conditioning units, central air is still rare in much of

Europe. In the United Kingdom and Germany, respectively, a 2022 industry estimate reports that only 3 percent of homes have air conditioning.[3] While this is largely due to the fact that, until recently, extreme heat has been a relatively rare phenomenon in most European countries, cultural preferences for fresh air also contribute to public resistance towards air conditioning. Still, the climate data firm Kayrros reports that the brutal 2022 heat wave caused a spike in AC demand. During the summer, AC sales rose by 1,420 percent in the UK, and energy consumption also skyrocketed.[4]

# Developing Nations

The greatest increases in AC consumption are expected to occur in China and the Global South. In their report on "The Future of Cooling," the International Energy Agency notes that most people in the world's hottest countries "have not yet purchased their first AC."[5] China, whose AC sales grew by a factor of

---

[3]Unsigned, "Air Conditioning in Europe," Inaba Denko, https://www.inaba -denko.com/en/inaba_note/detail/1 (June 3, 2022).
[4]Rosie Frost, "'Brutal' Heatwaves Across Europe Create Vicious Cycle of More Air Con and Higher Emissions," https://www.euronews.com/green /2022/07/19/brutal-heatwaves-across-europe-create-vicious-cycle-of-more -air-con-and-higher-emissions.
[5]Brian Dean, John Dulac et al., "The Future of Cooling: Opportunities for Energy-Efficient Air Conditioning," International Energy Agency (May 2018) https://www.iea.org/reports/the-future-of-cooling.

5 between 2000 and 2017, recently surpassed the US as the nation with the most AC units installed—China has 569 million, compared with 374 million in the US and 243 million in the entire rest of the world.[6] And there is still tremendous capacity for growth, since forty percent of households in China do not own an air conditioner. As the critic Ben Tran has pointed out in the context of contemporary Vietnam, the uneven spread of private air-conditioned spaces is especially striking in officially socialist nations like Vietnam and China, where separation from the shared, communal air seems "incommensurate with the envisioned socialist nation-state."[7]

While China has made important shifts toward renewable energy sources, the 2022 heat wave in Sichuan—combined with a drought that reduced hydropower generation by half—demonstrates how surges in AC usage can challenge sustainable energy infrastructures. With its power grid pushed beyond capacity by business and home air conditioners, Sichuan turned to its coal power plants to help meet the unprecedented demand for energy.[8]

---

[6]Chiara Delmastro, Yang Zhang et al., "The Future of Cooling in China," International Energy Agency (June 2019) https://www.iea.org/reports/the-future-of-cooling-in-china.
[7]Ben Tran, "Air-Conditioned Socialism: The Atmospheres of War and Globalization in Lê Minh Khuê's Fiction," *Cultural Critique* 105 (Fall 2019), 108.
[8]Nectar Gan, "China's Worst Heat Wave on Record is Crippling Power Supplies," *CNN.com* (Aug 26, 2022) https://www.cnn.com/2022/08/26/china/china-sichuan-power-crunch-climate-change-mic-intl-hnk/index.html.

The International Energy Agency projects that "By 2050, around 2/3 of the world's households could have an air conditioner. China, India, and Indonesia will together account for half of the total number."[9] While increasing AC consumption in so-called "developing countries" will likely have considerable effects on greenhouse gas emissions, it is important to keep in mind that the distribution of AC access is inequitable—and that it exacerbates existing inequalities. For example, the high-end Siam Paragon shopping mall in Bangkok "consumes nearly twice as much power annually as all of Thailand's underdeveloped Mae Hong Son Province, home to about 250,000 people."[10] To feed consumer demand for thermal comfort in one of the world's hottest cities, Thailand is importing hydropower from Laos, where megadams have disrupted and displaced river environments and fishing communities. As Thailand plans more megadam projects, critics point out that they are likely to "devastate fish populations, harm agriculture, and hurt culture and tourism—threatening the livelihoods of the nearly 65 million people who rely on the river for income and food."[11]

---

[9]Dean, Dulac, et al, "The Future of Cooling."

[10]Adam Pasick, "Bangkok's Lavish, Air-Conditioned Malls Consume as Much Power as Entire Provinces," *Quartz* (Apr 6, 2015) https://qz.com/376125/bangkoks-lavish-malls-consume-as-much-power-as-entire-provinces/

[11]Unsigned, "Gigawatts for Megaspenders," *Mekong Eye* (Sep 5, 2016) https://www.mekongeye.com/2016/09/05/gigawatts-for-gigaspenders-infographic-shows-bangkoks-luxury-malls-use-more-energy-than-some-provinces/.

Some of the most widely publicized air conditioning projects have occurred in the Middle East, where resource-rich countries can afford to spend billions on mitigating the effects of climate change in one of the world's hottest regions. Drawing on its vast oil reserves, Dubai has created "air-conditioned and cooled climatic biospheres of luxury." Developers there have begun construction on The Mall of the World, an air-conditioned domed city covering over 1,100 acres that would include districts dedicated to culture, hospitality, shopping, and wellness. As the geographer Peter Adey points out, however, Dubai's air-conditioned consumer environments, built on a finite resource, are "likely to dissolve back into the sands of the desert once its economy has run dry of oil."[12] Meanwhile, Qatar—a rapidly urbanizing country where average temperatures have increased by more than 3.6°F since the nineteenth century—has been experimenting with air conditioning outdoor spaces like pedestrian walkways, restaurant patios, and stadiums.

Amid media reports about increasing AC usage in developing countries and extravagant AC projects in oil-rich nations, it is important to keep in mind the historically high energy usage of the US documented throughout this book. As Jane Hu observes, the outraged response to Qatar's outdoor air conditioning in Western media conveniently overlooked the fact that North American cities like Phoenix

---

[12]Peter Adey, *Air* (London: Reaktion, 2015), 156, 157.

had already been using outdoor air conditioning for some years, and that Qatar is responsible for less than 1 percent of global emissions—just 0.01 percent of US emissions in 2014.[13] Even in the Middle East, the US has been responsible for outlandish fuel expenditures in the name of thermal comfort. Of the vast amounts of diesel fuel that the US military hauled into bases in Iraq and Afghanistan, "fully 85 percent was going to power air conditioning systems via generators." To enhance the morale of military personnel fighting a war largely motivated by oil, US bases offered a range of climate-controlled amenities including department stores, name brand fast food restaurants, and movie theaters.[14] Like Dubai, the US military built what amounted to a vast and unsustainable air-conditioned city in the torrid desert—a city dedicated to war-making and securing access to cheap oil.

If the adoption of air conditioning has been uneven across the globe, so have its effects on local ecologies and public health. Air conditioners and heaters have historically relied heavily on electricity generated from coal—a fuel whose extraction and combustion have devastating effects on land, air, water, and health. Even as wealthier countries

---

[13] Jane Hu, "Qatar's Outdoor Air Conditioning is Not the Real Climate Villain," *Slate* (Nov 4, 2019), https://slate.com/technology/2019/11/qatar -outdoor-air-conditioning-climate-change-emissions.html.
[14] Stan Cox, "Militarism, Torture . . . and Air Conditioning?"

transition to cleaner and renewable energy sources, India, Pakistan, and China have turned to coal as a way to meet energy demand, especially during dangerous heatwaves. As one energy researcher in India explains, "today's warming is on account of what the West has done over the last 150 years, which means that if the only option to give [people] cooling right now is through the use of air conditioners and fans that run on electricity that unfortunately has to be provided from coal, so be it, because this is here and now danger."[15]

As the example of Laos's megadams illustrates, even "cleaner" energy sources like hydropower (generated by environmentally destructive dams) or natural gas (often extracted through fracking) can put environments and their human and nonhuman inhabitants at risk. Worldwide, megadams and fuel extraction disproportionately affect Indigenous communities who rely on the land for physical survival and cultural continuance. Even the thermally conductive coils found in most heating and cooling technologies are made of copper—a metal typically extracted through ecologically destructive open pit mining, and often on Indigenous ancestral lands such as those of the Blaan people in the Philippines, the Ayamara Pakajaqi nation in

---

[15]Benjamin Storrow, Sara Schonhardt, "Sweltering India Turns to Superheating Coal for Cooling," *Scientific American* (June 2, 2022), https://www.scientificamerican.com/article/sweltering-india-turns-to-superheating-coal-for-cooling/.

Peru, and the Apache Nation on Turtle Island. As Lauren Redniss notes in her graphic nonfiction book about the San Carlos Apache Indian Reservation's fight against a planned copper mine, the transition towards more sustainable technologies will "almost certainly cause a further surge in demand for copper," because its high conductivity makes it indispensable to many clean technologies.[16]

Globally, air conditioning provides comfort to those who can afford it. But does it provide safety for those who are most vulnerable to extreme heat? For poor communities in hot, developing countries, air conditioning remains inaccessible even as its emissions have contributed to increasingly deadly heat events such as the 2015 heat wave that killed around 4,500 people in India and Pakistan. Among adults over sixty-five worldwide, heat-related deaths rose by eighty-one percent between the early 2000s and 2019. Despite the country's significant recent increases in AC usage, one study found that "the number of deaths caused by heatwaves in China has increased rapidly since 1979, from 3,679 persons per year in the 1980s to 15,500 persons per year in the 2010s."[17] A recent collaborative publication by the UN Office for the Coordination of Humanitarian Affairs and the International

---

[16]Lauren Redniss, *Oak Flat: A Fight for Sacred Land in the American West* (New York: Random House, 2020), 63.
[17]Li Yuan, "Mortality Caused by Heatwaves in China has Increased Since 1979," *phys.org* (July 21, 2022), https://phys.org/news/2022-07-mortality -heatwaves-china.html.

Federation of Red Cross and Red Crescent Societies projects "staggeringly high" death rates from extreme heat, which by the end of the century will be comparable to death rates caused by "all cancers or all infectious diseases." This report documents the inequitable distribution of heat vulnerability, noting that the greatest increases of vulnerable populations living in extreme heat are projected to occur in West Africa and Southeast Asia, and that heatwaves "could bring recurring life-threatening conditions to up to 600 million people" in parts of North Africa and the Middle East.[18] In light of such projections, it seems more important than ever to consider how air conditioning can be made available on equitable terms as a technology of public health, especially in hot regions most impacted by colonialism and climate change.

## "Air-Conditioned Coffins"

Despite the fact that air conditioning can help protect some people from extreme heat exposure, many people in colonized and postcolonial nations perceive AC as a colonial imposition.

---

[18]*Extreme Heat: Preparing for the Heatwaves of the Future* (UN Office for the Coordination of Humanitarian Affairs, International Federation of Red Cross and Red Crescent Societies, and the Red Cross Red Crescent Climate Centre, 2022), https://www.ifrc.org/sites/default/files/2022-10/Extreme -Heat-Report-IFRC-OCHA-2022.pdf.

In his influential manifesto "Towards a New Oceania" (1976), the Samoan writer Albert Wendt denounced the modern hotel rooms that were being built throughout Oceania (the Pacific Islands) as "air-conditioned coffins lodged in air-conditioned mausoleums." Wendt warned against architecture as another mode of Western invasion:

> [I]ts most nightmarish form is the new type tourist hotel—a multi-storied edifice of concrete/steel/chromium/ and air conditioning. This species of architecture is an embodiment of those bourgeois values I find unhealthy/ soul-destroying: the cultivation/worship of mediocrity, a quest for a meaningless and precarious security based on material possessions, a deep-rooted fear of dirt and all things rich in our cultures, a fear of death revealed in an almost paranoic quest for a super-hygienic cleanliness and godliness.[19]

Instead of the environmental connectedness and passive cooling afforded by the traditional Samoan *fale* (a thatched hut supported by columns with no walls), air-conditioned buildings were "constructed of dead materials." Wendt insisted that the "deodorized/sanitized comfort" offered by air conditioning was desensitizing and soul-killing for Oceanic people and traditions.

---

[19] Albert Wendt, "Towards a New Oceania," *MANA Review* 1:1 (1976), 57, 56.

Wendt's arguments are an important reminder that, in addition to making a broad range of climates more comfortable and valuable for imperialist functionaries, colonial settlers, and tourists, architecture itself does the work of colonialism by replacing place-based traditions and knowledges with a neutralized and environmentally destructive indoor environment that, thanks to air conditioning, could be reproduced anywhere on the planet. Here it is helpful consider the question that Hi'ilei Julia Hobart poses in her book about cold infrastructures in Hawai'i: "What are the implicit understandings of environment, climate, and embodiment that underpin human needs and desires for the cold within the tropics?"[20] Air conditioning doesn't just remake the air—it quietly remakes conceptions of the human and ways of relating to the environment.

One way that air-conditioned comfort "kills the soul" is by supplanting local, climate-appropriate architectures like the Samoan *fale* or the thermally efficient adobe structures built by some Indigenous nations across the Americas. In North Africa and West Asia, wind towers that had used prevailing winds to cool and ventilate buildings since ancient times were largely neglected by architects following the introduction of air conditioning in the twentieth century. (In the twenty-

---

[20]Hi'ilei Julia Kawehipuaakahaopulani Hobart, *Cooling the Tropics: Ice, Indigeneity, and Hawaiian Refreshment* (Durham: Duke University Press, 2023), 3.

first century, there has been a resurgence of interest in wind towers as a model for sustainable architecture.) The siesta—a practice of thermal regulation that enables people to rest and cool down during the hottest part of the day—was already in decline when Mexico banned it in government offices in 1999. Spain also effectively outlawed the practice (associated with a three-hour midday break) when it shortened the workday in 2016. Climate-controlled eating spaces detach foods like gazpacho, melons, warming spices, or hot pot from their traditional functions of thermal regulation. Local traditions—often connected with natural and seasonal cycles—are eroded as air conditioning homogenizes people's senses of temperature, place, and time.

## "The Great Uncooled"

Stories about the unsettling effects of economic modernization and "development" in the Global South can shed light on how new cooling infrastructures feel on the ground for people from different cultural and socioeconomic backgrounds. *Moth Smoke* (2000), the debut novel by the British Pakistani author Mohsin Hamid, depicts the devastating effects of globalization on everyday life in Lahore: along with widespread unemployment, currency devaluation, increasingly corrupt institutions, weakened infrastructure, and a widening wealth gap, Pakistan was increasingly isolated internationally as a result of its nuclear test conducted in May 1998 in response to India's nuclear tests. Hamid singles out the electric grid—and, in particular, the availability of air conditioning—as a powerful infrastructure that unevenly distributes the everyday, embodied effects of growing socioeconomic disparities. The novel's anti-hero, Daru, has his electricity cut off soon after losing his job as a bank clerk. As a result, the wealth and privilege that separate him from his elite friends take the form of not only economic and cultural distinctions, but embodied differences: without air conditioning, Daru sweats, stinks, and stews with rage and disaffection.

In Lahore—where regular summer temperatures over 104°F coexist with frequent power outages and widespread poverty—air conditioning is both a visible status symbol and a powerful determinant of life possibilities. In a chapter titled "What Lovely Weather We're Having (or the Importance of Air-conditioning)," Hamid puts the novel's plot on hold to reflect on how uneven access and attitudes concerning air conditioning have influenced the novel's characters and key events. The chapter is built around a lecture in which a (fictional) economics professor asserts,

> There are two social classes in Pakistan. [. . .] The first group, large and sweaty, contains those referred to as the masses. The second group is much smaller, but its members exercise vastly greater control over their immediate environment and are collectively termed the elite. The distinction between members of these two

groups is made on the basis of control of an important resource: air conditioning.[21]

Hamid goes on to suggest that key events in the novel—from the accidental death of Daru's mother (struck by a stray bullet when sleeping on a balcony to stay cool during a power outage) to his affair with his friend's wife (whose marital problems stemmed from different attitudes toward AC)—were the outcomes of uneven access to AC. We learn that another character, who grew up living on the street, spent much of his life believing that air conditioners produce hot air, having only experienced them from the outside. This character's lack of knowledge concerning air conditioning as a cooling technology underscores the privilege of those on the inside, who might be just as unaware of the heat that their AC condensers are releasing onto the streets outside.

*Moth Smoke* invites readers to consider not only how air conditioning and different ideas about thermal comfort affect us in subtle but powerful ways, but also the ethical implications of an infrastructure that cools indoor air by releasing waste heat into spaces and populations located outside the bubble. By situating the entire chapter on AC as part of a criminal trial, *Moth Smoke* asks whether and how the air—and uneven access to cooling technologies—should be accounted for in the law.

---

[21]Mohsin Hamid, *Moth Smoke* (New York: Picador, 2000), 102.

The novel also shows how Lahore's extreme infrastructural, economic, and thermal disparities have been intensified by Pakistan's entanglements with the global economy. As Lahore's air-conditioned class—a "mixed lot [of] Punjabis and Pathans, Sindhis and Baluch, smugglers, mullahs, soldiers, industrialists"[22]—gain access to lives spent in climate-controlled bubbles that mark their status as global elites, their complicity with government corruption and investments in privatized infrastructures (including electric generators) leave everyone else increasingly exposed to heat. *Moth Smoke* approaches air conditioning as an occasion for heightened narrative realism—for telling stories that position human actions as the products, in part, of the hidden influences of air conditioning and climate change.

*Air Conditioner* (*Ar Condicionado;* Fig. 7), a 2020 film by the Angolan screenwriter and director Fradique (Mário Bastos), takes a more speculative approach to depicting the uneven distribution of air conditioning technologies in the Global South. In a magical realist twist, the film's premise is that window AC units in Luanda—the capital of Angola— have inexplicably begun falling onto the streets below. But this bizarre epidemic does not erase the distinction between air-conditioned elites and the "great uncooled." Instead, falling air conditioners are killing people on the streets, contributing to a "death toll linked to heat and air conditioner collapse

---

[22]Ibid., 103.

throughout the country." The film follows a housemaid and security guard who have been ordered to get their employer's window unit reinstalled, despite official recommendations that existing AC units be removed for public safety.

By focusing on air conditioning, Fradique foregrounds a powerful symbol of wealth and status for a middle class emerging amid the social, infrastructural, and ecological devastation left in the wake of Angola's War of Independence from Portugal (1961-75) and protracted Civil War (1975-2002). In Angola's hot and crowded cities, air conditioning enables middle class residents to distance themselves from urban smells, and to wear perfumes as markers of status.[23] In the film, a news commentator suggests that falling ACs epitomize the social gap between those who "suffer" and elites who "live it up . . . in utter comfort while common folk perish." *Air Conditioner* dwells on the run-down spaces where Luanda's invisible workers struggle and support one another: alleys, streets, shops, and the apartment building that Zezinha and Matacedo are employed to maintain and protect. Air conditioning is not just a luxury enjoyed by the rich—everyone feels its effects. In one scene, we hear about a 93-year-old who died after tripping over a broken air conditioner; in another, people fight for the right to salvage air conditioners as they're falling from above. Meanwhile,

---

[23]Jess Auerbach, *From Water to Wine: Becoming Middle Class in Angola* (Toronto: University of Toronto Press, 2020), 34-42.

Zezinha and Matacedo's boss doesn't mince words about the fact that he values his apartment's AC unit far more than the workers he charges with repairing it.

*Air Conditioner* presents Angola's ongoing and uneven adoption of artificial cooling as part of a broader social transformation. As the film begins, Zezinha recalls her father's life as a fisherman before their family was evicted and migrated to the city—how he used to watch the sea until the temperature was just right. "He used to say, 'the wind is the almighty.' I'm talking about the real wind, not the air conditioner's breeze." Mr. Mino, the mystic electrician who repairs the AC unit, notes that all the trees in the city have been cut down—the only plants left in Luanda are the ones in his shop's back room. Not only the air and the city's mood but people's entire way of relating to their environment have been quietly reshaped by AC. As a hip-hop song played in the film comments, "They conditioned everything, even the air we breathe." *Air Conditioner* points toward two possible solutions to this predicament. First, a news commentator proposes a political fix: "We shall try to implement a social housing policy that is in harmony with the weather conditions of our country." Later, Mr. Mino experiments with a more imaginative way of repurposing modern technologies, cobbling AC equipment and other electronics into machines that visualize working people's memories and dreams. Instead of cutting Luanda's people off from their collective memories of place, weather, trees, and social relationships, Mr. Mino refurbishes AC parts into a tool for re-experiencing and renewing ways of life that are being displaced by the drive towards modernization.

Fradique's film dramatizes air conditioning—a technology integral to class identity, contemporary architecture, and urban planning—as an "unsustainable" infrastructure that literally cannot be held up or sustained. While the film's air conditioners falling from apartment windows injure and kill people on the streets below, they also provide promising opportunities for salvage—transforming AC parts into machines for accessing memories and dreams that might inspire a future where cool, breathable air is not monopolized in bourgeois bubbles.

Stories like *Moth Smoke* and *Air Conditioner* offer important counterpoints to the narratives of blame (even blame for what hasn't happened yet, as in "air conditioning in X country is going to increase dramatically by 2050") and victimization that often orient discussions of air conditioning in the Global South. Hamid and Fradique complicate our understandings of global cooling by emphasizing how AC intensifies social divides, transforms understandings of human agency, and even opens up possibilities for imagining and designing more livable futures.

# CONCLUSION

# BURSTING OUR BUBBLES

In his 2015 encyclical on environmental protection, *Laudato Si'* Pope Francis identifies air conditioning as an exemplary instance of the entrenched "habits of consumption" that are simultaneously damaging the planet and the global poor. Commenting on capitalism's production of ever greater demand for AC, he writes: "An outsider looking at our world would be amazed at such behavior, which at times appears self-destructive."[1] Here, the Pope (whose environmental views have been far more progressive than his comments on gender and abortion) imagines how our contemporary patterns of air conditioning consumption would look from

---

[1]Pope Francis, *Laudato Si': On Care For Our Common Home* (2015), 41, https://www.vatican.va/content/dam/francesco/pdf/encyclicals/documents/papa-francesco_20150524_enciclica-laudato-si_en.pdf.

an extraterrestrial, God's-eye perspective. Seen impartially, it seems undeniable that we are using air conditioning in unsustainable and destructive ways.

The problem, however, is that there is no impartial, "outsider" perspective on air conditioning. We're all conditioned by its emissions and expenditures, whether we use it or not. Just as significant portions of this book were written in air-conditioned spaces, the Pope (as critics were quick to point out) probably composed *Laudato Si'* in his air-conditioned residence; and his encyclical itself is likely to be preserved for posterity in a climate-controlled archive. The point here isn't to accuse air conditioning's critics of hypocrisy, but to highlight how deeply it is entrenched into many people's daily lives and habits; even if I avoided AC at home, it would be impossible to avoid it in the libraries, archives, university classrooms, conference centers, and cafes where I do much of my work. Similarly, chances are good that you are reading this book in a climate-controlled space, or would be if the weather weren't already comfortable outside. However self-destructive AC might appear to an outsider, from the inside it tends to feel comfortable, "normal"— even unavoidable. With time, for those on the inside, AC feels like nothing at all—just the absence of discomfort. Air conditioning immerses us in a fantasy of disembodiment—a space in which our relationships with temperature and weather seem immaterial.

This is another lesson that air conditioning teaches us: our embodied sense of comfort—along with all the ideas and values wrapped up with it—is not innate, but constructed through historical processes, marketing campaigns, and constant recalibration. This is why thermal comfort zones and the desirability of air conditioning vary considerably across cultures. If we can be conditioned to feel that a constant, comfortable ambient temperature is a non-negotiable human need, we can also be reconditioned to relinquish this deeply embodied demand for comfort. What would it take to undo the effects of air conditioning on our embodied habits and predispositions? How can we shift our relationship to AC infrastructures that have quietly and continuously been shaping us—and our planet—all along?

To undo the desensitizing effects of air conditioning, we need to inhabit the different ways of sensing, thinking, and acting explored by many of the writers and artists discussed in this book. Senses like thermoception and smell, often undervalued in a culture oriented by the relatively disembodied sense of vision, can enable us to experience how deeply we're enmeshed with the world. Instead of relating to objects in a vacuum, these senses entangle us with climate and atmosphere materially—even metabolically. Where climate control privileges visual apprehension (for example, of a summer blockbuster in a movie theater, clean-looking white furniture in a suburban home, or well-preserved artworks in a museum), variable temperatures align the speed of molecular reactions in our bodies with the speed of molecules in the air around us. Discomfort can be an important condition of knowledge. Temperature attunes our metabolism to the time of day, to the weather, and to the ways in which human activity has been transforming the climate. Attending to our thermal sensations—including uncomfortable ones—can induce us to explore a somewhat broader temperature range than prescribed "comfort zones" allow, not just as an undesirable thermal condition that should be avoided, but as a way of shifting how we inhabit and relate to the world. Instead of driving us into private thermostatic bubbles, the sense of temperature can help interconnect humans and non-humans across a range of atmospheric conditions.

Becoming conversant with a broader range of temperatures would decenter thermostatic comfort as a *sine qua non* of bourgeois existence. Expanding the range of temperatures we spend time in can recalibrate our comfort zones, in turn reducing our dependence on air conditioning and making more space for diverse practices of thermal regulation. Unsettling how we think about comfort is crucial to working out a more equitable and sustainable approach to air conditioning, because the distribution of air conditioning should be determined not by private comfort, but by public health. As extreme weather becomes increasingly common and widespread, the thermal preconditions for health should be acknowledged as a human right. Researchers should focus not only on temperature's effects on comfort and "productivity" (defined in culturally specific ways), but also on the often invisible, indirect ways in which temperature affects physical and mental health.

Public cooling and warming centers that provide thermal comfort much more equitably and sustainably than private spaces should be widespread, accessible, and inclusive. Air conditioning should be accessible in the commons, because its environmental impacts are already distributed across the entire planetary commons.

Is it possible to design thermally habitable spaces that refuse to imagine comfort only as an attribute of private, thermally standardized bubbles? In a project entitled *Thermal Justice*, the architect Amber Godfrey explores how local vernacular designs might offer a more accessible alternative to the bubble. One component of this work, "The Temple of Thermal Diversity," envisions a pavilion consisting of a range of thermal zones created by passive air flow. The idea of thermal diversity builds on Philippe Rahm's earlier architectural experiments with sustainable buildings designed to take advantage of convection, where passive air circulation creates different microclimates within a home or office building attuned to appropriate practices and activities: for example, rooms for sleeping, exercising, bathing, or moving around the kitchen while wearing heavier or lighter clothing. For both Godfrey and Rahm, diverse thermal zones hold space for a broad and open-ended range of possible activities, feelings, and sensations.

Another component of Godfrey's project, "The Parliament of Thermal Justice" focuses on Jamaica (where Godfrey grew up)—a tropical island where, "more and more, developers build buildings unable to be passively comfortable in the name of cost efficiency and 'modern architecture,' and tenants live with the exceedingly high cost of being cool." This high cost, coupled with the socioeconomic legacies of slavery and colonialism in Jamaica, means that both air conditioning access and exposure to air conditioning's environmental externalities are unevenly distributed across racial lines. Responding to a design competition for a new Parliament building, Godfrey envisions a porous structure based on vernacular, climatically appropriate architectural elements. What kind of legislation and governance would emerge from a parliament that shares its atmosphere with the people it represents? The implicit argument of Godfrey's work is that passive cooling, a diversity of thermal zones, and "dissolving the barrier between interior and exterior" will make space for a more open and environmentally equitable society.[2] Whereas corporations, governments, and public institutions tend to approach the ideas of diversity and equity in representational (and often superficial) terms, *Thermal Justice* invites us to consider how these values might be enacted materially in the atmospheres we design and inhabit.

Along with many of the other artworks considered in this book, *Thermal Justice* offers an important counterpoint to the argument that AC overconsumption will be fixed by more sustainable, efficient cooling technologies and the

---

[2]Amber Godfrey, "Thermal Justice," *KoozArch* (Jul 23, 2021), https://koozarch.com/interviews/thermal-justice.

transition to renewable energy sources. As Pope Francis writes, "Technology, which, linked to business interests, is presented as the only way of solving these problems, in fact proves incapable of seeing the mysterious network of relations between things and so sometimes solves one problem only to create others."[3] To become reacquainted with the "mysterious network of relations" that has been cut up into sealed-off bubbles, we will have to re-sensitize ourselves to the common (and not always comfortable) atmosphere outside, to the ongoing interactions between interior and exterior climates, and to the processes of thermal injustice that shape our world.

In his stove-heated room, Descartes imagined the separation of mind and body on the basis of thermal comfort. What if, instead, we took thermal variability and discomfort—and our unavoidable enmeshment with the air and climate—as the basis for thought and action? I hope this book will make a small contribution to understanding and unraveling the profound ways in which air conditioning has *conditioned* the ways that many of us sense, interpret, and interact with the world, and that recognizing our status as beings entangled with air and climate will help us realize more livable and equitable ways of engaging with the atmospheric commons.

---

[3]Pope Francis, 41.

# ACKNOWLEDGMENTS

This book was written on unceded lands that have long been inhabited and stewarded by the Kanien'kehá:ka, Patwin, Nisenan, Southern Maidu, Valley and Plains Miwok people.

This book has benefited from conversations and exchanges about AC, climate, comfort, and the senses I was fortunate to have with David Howes, Andrew Kettler, Edlie Wong, Mike Ziser, Tobias Menely, Sophia Bamert, Kristin George Bagdanov, Matthew Stratton, Erica Fretwell, Will Tullett, Jean-Thomas Tremblay, Elizabeth Miller, Margaret Ronda, Kathleen Frederickson, Emily Yeh, Bharat Venkat, the wonderful students in my Sensory Ecologies class, and the colleagues who generously engaged with my presentations at UCLA, Colby College, National Chung Hsing University, and Uncommon Senses III. I'm especially grateful to Christopher Schaberg, Akua Banful, Kara Murphy Schlichting, and Rachael Dewitt, who provided brilliant and generous feedback on drafts. Haaris Naqvi and Rachel Moore at Bloomsbury provided expert editorial input, and the series editors Ian Bogost and Christopher Schaberg have been wonderfully supportive. Mistakes and missteps are all mine.

I wouldn't be around to write this book without the love and support of Kang, Hsiang-Lin, Kang, Lin, Cristina, Jason, Kile, Kalissa, Myron, Ellie, and Lucy. This book is dedicated to Beenash and Zayn; I'm lucky to share the air with you.

Some portions of Chapter 5 draw on research first published in my article, "Race, Urban Heat, and the Aesthetics of Thermoception," *American Literary History* 35:2 (Summer 2023) 769-94.

# INDEX